More Praise for

The Spark of Life

"A rich and stimulating account of how electricity and the varied ways in which animals and plants produce it explain so much of evolutionary biology."   —*Wall Street Journal*

"You know you're in for an interesting read when a book opens with a cautionary note warning you not to electrocute yourself (or any animals) if you happen to re-create the experiments within. . . . [Ashcroft] keeps things moving at a colloquial clip."   —*Daily Beast*

"Lively, conversational prose—refreshingly accessible to any lay reader prepared to make a little effort. . . . [A] positively charged little book."   —*Telegraph*

"With style and enthusiasm, Oxford professor Ashcroft reveals the ubiquitous role electricity plays in our bodies. . . . Ashcroft's writing is clear and accessible, offering surprising insights into the 'electrical machine' we call the human body."   —*Publishers Weekly*

"[Ashcroft] communicates complex science with engaging passion and eloquence."   —*Observer*

"The human body is a symphony of complicated chemical and electrical signals. Let Ashcroft's book serve as your program guide."
                                                        —*Booklist*, starred review

"Very well researched. . . . But this is not simply a book for a niche market; it should also be well received by a wide audience. It offers a fascinating and accessible introduction to what is often perceived to be a difficult subject. . . . Yet all this science is explained while maintaining the reader's interest with a human narrative."
—*Times Higher Education*

"Ashcroft clearly and patiently introduces complicated science while enlivening her narrative with fascinating tales of electric eels, fugu fish poisoning, and fainting goats." —*Library Journal*

"Informative and relatable real-life anecdotes tie many of [the] scientific concepts together and provide tantalizing hints of what new applications of electricity may be on the horizon. . . . A captivating read sure to pique the interest of any science fan." —*Kirkus Reviews*

"An extraordinary fusion of culture and cutting-edge science that has something of the 'watcher of the skies' about it."
—*Engineering and Technology*

"Ashcroft covers questions as far ranging as why an electric eel doesn't shock itself to how LSD works. . . . That she does it with scientific clarity is to be expected; that she does it with pylon sized chunks of wit is a joy to discover." —*MoneyMaker*,
Book of the Week

"If you love science and are curious about what makes your body work, then you should read this book." —The Skeptic's Dictionary

The Spark of Life

Electricity in the Human Body

FRANCES ASHCROFT

Line drawings by Ronan Mahon

W. W. Norton & Company

New York · London

For Rosy and Charles

Copyright © 2012 by Frances Ashcroft
Line drawings copyright © 2012 by Ronan Mahon
First American Edition 2012

For information about permission to reproduce selections from this book, write to
Permissions, W. W. Norton & Company, Inc., 500 Fifth Avenue, New York, NY 10110

For information about special discounts for bulk purchases, please contact
W. W. Norton Special Sales at specialsales@wwnorton.com or 800-233-4830

Manufacturing by RR Donnelley, Harrisonburg
Production manager: Devon Zahn

Library of Congress Cataloging-in-Publication Data

Ashcroft, Frances M.

The spark of life : electricity in the human body / Frances Ashcroft ;
line drawings by Ronan Mahon. — 1st American ed.
p. cm.
Includes bibliographical references and index.
ISBN 978-0-393-07803-9 (hardcover)
1. Electrophysiology—Popular works. 2. Human physiology. I. Title.
QP341.A77 2012
612'.01427—dc23
2012021264

ISBN 978-0-393-34679-4 pbk.

W. W. Norton & Company, Inc., 500 Fifth Avenue, New York, N.Y. 10110
www.wwnorton.com

W. W. Norton & Company Ltd., Castle House, 75/76 Wells Street, London W1T 3QT

1 2 3 4 5 6 7 8 9 0

'Man is no more than electrified clay'

Percy Bysshe Shelley

Contents

Cautionary Note

Electricity is highly dangerous when used incorrectly. The reader is strongly advised *not* to try any of the experiments described in this book on themselves, on others, or on animals. The author and publishers accept no responsibility for any resulting harm if the reader should fail to heed this advice.

Introduction
I Sing the Body Electric

Then felt I like some watcher of the skies
When a new planet swims into his ken;
Or like stout Cortez when with eagle eyes
He star'd at the Pacific – and all his men
Look'd at each other with a wild surmise –
Silent, upon a peak in Darien.

John Keats, 'On first looking into Chapman's Homer'

When he was just a few months old, James suddenly developed diabetes so severe that he required hospitalization. He faced a lifetime of insulin injections. Over the course of the next few months it became clear that he was also developing more slowly than most children and that he had problems with walking and talking. By the time he was five years old, he had only just started to walk, he was still unable to communicate and he had the temper tantrums of a two-year-old. Life was far from easy for his anxious parents.

It was later discovered that James has a very rare form of diabetes caused by a genetic defect (a mutation) in a protein known as the K_{ATP} channel that is important for both insulin secretion and brain function. Some mutations in the K_{ATP} channel simply cause diabetes, but around 20 per cent of them, including the one James has, also produce a constellation of neurological difficulties, including developmental delay, hyperactivity, behavioural problems and muscle dysfunction. All of these symptoms arise because the K_{ATP} channel influences the electrical activity of the insulin-secreting cells as well as that of the muscles and brain. It turns out that James's story is closely entwined with my own, for the K_{ATP} channel has been my life's work and understanding how it operates has enabled

James to replace the multiple daily insulin injections he once needed to control his diabetes with just a few pills.

Diabetes occurs when the beta-cells of the pancreas do not release enough insulin for the body's needs, so that the blood sugar level rises. Back in 1984, I discovered that the K_{ATP} channel sits in the membrane that envelops the beta-cell and regulates its electrical activity and thereby insulin release. The channel functions as a tiny, molecular pore that is indirectly opened and closed by changes in the blood sugar concentration: when the pore is closed insulin secretion is stimulated and when it is open insulin release is inhibited.[1]

I vividly remember the day I made that discovery. As so often happens, the breakthrough came late at night when I was working alone. I had hypothesized that adding glucose to the solution bathing the beta-cells would cause the channel to shut. Yet when it did, I felt certain it must be a technical error. So certain, in fact, that I almost ended the experiment. But just in case I was wrong, I tested the effect of removing the sugar, reasoning that if glucose were indeed regulating the channel activity its removal should cause the pore to reopen, whereas if it were simply a technical problem the channel should remain closed. After several agonizingly long minutes the channel opened once again. I was ecstatic. I was dancing in the air, shot high into the sky on the rocket of excitement with the stars exploding in vivid colours all around me. Even recalling that moment sends excitement fizzing through my veins, and puts a smile on my face. There is nothing – nothing at all – that compares to the exhilaration of discovery, of being the first person on the planet to see something new and understand what it means. It comes all too rarely to a scientist, perhaps just once in a lifetime, and usually requires years of hard grind to get there. But the delight of discovery is truly magical, a life-transforming event that keeps you at the bench even when times are tough. It makes science an addictive pursuit.

That night I felt like stout Cortez, silent upon his peak in Darien, gazing out across not the Pacific Ocean, but a landscape of the mind. It was crystal clear where my mental journey must take me, what experiments were needed and what the implications were. Next morning, of course, all certainty swept away, I felt sure my beautiful result was merely a mistake. There was only one way to find out. Repeat the experiment – again and again and again. That is the daily drudgery of a scientific life: it is very far from the ecstasy of discovery.

Even all those years ago, it was obvious that if the channel failed to close when the blood glucose level rose, insulin secretion would be prevented and the result would be diabetes. To prove it, we needed to find mutations in the DNA sequence that encodes the K_{ATP} channel protein in people with diabetes. It took ten years of work by many people throughout the world to identify that DNA sequence, and when we finally screened it for mutations we found . . . nothing!

It was my friend Andrew Hattersley who eventually found the first mutations, another ten years later. Andrew is a very special person. Tall, slim and sandy-haired, with an incisive mind and a warm compassionate nature, he is both a wonderful doctor and a brilliant scientist. He not only recognized that the mutations we were seeking would be more likely to be found in people who were born with diabetes (rather than those who developed it later in life); he also instigated a worldwide search to find them. When he and his associate Anna Gloyn identified the first mutation in 2003, he phoned me and invited us to collaborate with him. It was a call I will never forget.

Working together, we showed that the K_{ATP} channel mutations cause diabetes because they lock the channel permanently open, preventing electrical activity and insulin secretion. Even more excitingly, we found that the defective channels can be shut by drugs known as sulphonylureas that have been safely used for more than fifty years to treat type 2 (adult-onset) diabetes and which we already knew closed normal K_{ATP} channels.

In the past, patients who were born with diabetes were treated with insulin injections, as their symptoms suggested they had an unusually early-onset form of type 1 (juvenile) diabetes. In this disease the beta-cells are destroyed by the body itself and lifelong insulin is essential. Thus James and others like him were not given drugs, but immediately started on insulin. Our research suggested that instead such patients could be treated with sulphonylurea tablets and to everyone's delight the new therapy not only worked, but actually worked much better than insulin. Over 90 per cent of people with neonatal diabetes have been able to make the switch.

It is a rare privilege for a research scientist to see one's work translated into clinical practice, and even rarer to meet the people whose lives have been affected, so I have been extremely fortunate. Words cannot convey the extraordinary emotional experience of meeting the children and families whom your work has helped. To have, for example, a pretty young teenager turn to you and say 'Thanks to you I can wear a dress'. 'Why?' I inquired, puzzled. 'Because,' she replied, 'I no longer need a skirt or trouser waistband from which to hang my insulin pump.' An insulin pump, I quickly appreciated, is something of a constraint. Dashing in and out of the waves in the summer sea is simply not possible – each time the pump must be removed and reattached – and its bulky shape ruins the line of figure-hugging clothes. Drug therapy obviates these problems and banishes painful injections. But it also has more important benefits. For reasons still unclear (but which we are of course exploring), sulphonylureas produce a far more stable blood glucose level than insulin. Dramatic fluctuations in blood sugar become a thing of the past, and hypoglycaemic attacks are much less frequent (and in some cases virtually vanish). Unexpectedly, the average blood sugar level also decreases so that the risk of diabetic complications (kidney disease, heart disease, blindness and amputations) is reduced.

People with neonatal diabetes, and their families, have hailed the new therapy as a miracle. But this is no miracle: it is merely science. It is knowledge of exactly how ion channels regulate the electrical activity of the pancreatic beta-cells, and thereby insulin secretion, that has made it possible for patients to throw away their needles and insulin pumps and switch to tablets. And it is only by understanding more clearly the mechanisms underlying the electrical activity of nerve and muscle cells that eventually it may be possible to develop better therapies for their neurological problems.

We're all familiar with the fact that machines are powered by electricity, but it's perhaps not so widely appreciated that the same is true of ourselves. Your ability to read and understand this page, to see and hear, to think and speak, to move your arms and legs – even your sense of self – is due to the electrical events taking place in the nerve cells in your brain and the muscle cells in your limbs. And that electrical activity is initiated and regulated by your ion channels. These little-known but crucially important proteins are found in every cell of our body and in those of every organism on Earth, and they regulate our lives from the moment of conception until we draw our last breath. Ion channels are truly the 'spark of life' for they govern every aspect of our behaviour. From the lashing of the sperm's tail to sexual attraction, the beating of our hearts, the craving for yet another chocolate, and the feel of the sun on your skin – everything is underpinned by ion channel activity. Not surprisingly, given their ubiquity and functional importance, a multitude of medicinal drugs work by regulating the activity of these minute molecular machines, and impaired ion channel function is responsible for many human and animal diseases. Pigs that shiver themselves to death, a herd of goats that falls over when startled, people with cystic fibrosis, epilepsy, heart arrhythmias or (as I know only too well myself) migraine – all of us are victims of channel dysfunction.

An unusual tribute to the scientists and philosophers who contributed to the discovery of electricity hangs in the Musée d'Art Moderne in Paris. A giant canvas known as 'La Fée Électricité', which measures 10 metres high and 60 metres long, it was commissioned by a Paris electricity company to decorate its Hall of Light at the 1937 world exhibition in Paris. It is the work of the French Fauvist painter Raoul Dufy, better known for his wonderful colourful depictions of boats, and it took him and two assistants four months to complete. The Electricity Fairy sails through the sky at the far left of the painting above some of the world's most famous landmarks, the Eiffel Tower, Big Ben and St Peter's Basilica in Rome among them. Behind her follow some 110 people connected with the development of electricity, from Ancient Greece to modern times. As time and the canvas progress, the landscape changes from scenes of rural idyll to steam trains, furnaces, the trappings of the industrial revolution and finally the giant pylons that support the power lines carrying electricity to the planet.

Dufy's magnificent painting celebrates the scientists and engineers who shaped the world we live in today – Archimedes, Ampere, Edison, Franklin, Faraday, Ohm and others. But there is also a less well-known gallery of scientists who are the scientific descendants of Galvani, the discoverer of animal electricity. To these men and women we owe the drugs and technologies that we take for granted in our modern hospitals and our knowledge of how our bodies work. This book tells their stories. It charts the development of our ideas about animal electricity and shows how these are intertwined with our growing understanding of electricity itself. It explains how the electrical activity of our bodies is generated, and tells the dramatic, fascinating and sometimes tragic stories of what happens when things go wrong. What happens when you have a heart attack? Can someone really die of fright? Why are some people unable to stand when they eat a banana? What does Botox actually do? How does an electric eel give you a shock? How do vampire bats sense their prey? Is the red I see the same as you do?

This book provides the answers to these and other questions. It explains how ion channels work and how they give rise to the electrical activity of our nerves and muscles. It shows how ion channels act as our windows on the world and how all our sensory experience – from listening to a Mozart quartet to judging where a tennis ball will fall – depends on their ability to translate sensory information into electrical signals that the brain can interpret. It explores what happens when we fall asleep, or lose consciousness, and it discusses how our increasing knowledge of the electrical activity of the brain has stimulated and illuminated the link between mind and brain.

In essence, this book is a detective story about a special kind of protein – the ion channel – that takes us from Ancient Greece to the forefront of scientific research today. It is very much a tale for today, as although the effects of static electricity and lightning on the body have been known for centuries, it is only in the last few decades that ion channels have been discovered, their functions unravelled and their beautiful, delicate, intricate structures seen by scientists for the first time. It is also a personal panegyric for my favourite proteins, which captured me as a young scientist and never let me go; they have been a consuming passion throughout my life. In Walt Whitman's wonderful words, my aim is to 'sing the body electric'.

1

The Age of Wonder

I am attacked by two very opposite sects – the scientists and the know-nothings. Both laugh at me – calling me 'the Frog's Dancing-Master', but I know that I have discovered one of the greatest Forces in Nature.

Luigi Galvani[1]

'It was on a dreary night of November, that I beheld the accomplishment of my toils. With an anxiety that almost amounted to agony, I collected the instruments of life around me, that I might infuse a spark of being into the lifeless thing that lay at my feet. It was already one in the morning; the rain pattered dismally against the panes, and my candle was nearly burnt out, when, by the glimmer of the half-extinguished light, I saw the dull yellow eye of the creature open; it breathed hard, and a convulsive motion agitated its limbs.' Thus did Victor Frankenstein, in Mary Shelley's great gothic novel *Frankenstein*, written in 1818, record his creation of a monstrous being.

It is widely believed that electricity, in the form of a lightning bolt, was used to waken Frankenstein's creature to life. But this is a misconception that probably originates with the famous 1931 film in which Boris Karloff played the monster. Shelley herself was far more circumspect and refers only to the 'instruments of life'. Nevertheless, the novel leads us to infer that electricity was used to instill the monster with 'a spark of being', for Frankenstein gives a dramatic description of a lightning strike he witnessed as a young man that blasted an ancient oak tree to smithereens; and on inquiring about the nature of lightning from his father, he was informed it was 'Electricity'. Shelley also uses her preface to make a marriage of physiology and electricity – 'perhaps a corpse would be reanimated; galvanism had given a token of such things'.

Indeed, both Mary and her lover Percy Bysshe Shelley took a keen interest in the emerging science of electricity and its effects on the human body. Percy was a particular enthusiast having experimented with electricity at Eton, at Oxford and even at home – his sister recounts how she was terrified of being 'placed hand-in-hand round the nursery table to be electrified'. Percy was eventually sent down from Oxford for his atheist views and in 1810, during the winter vacation before his last term, he wrote to his tutor that he supposed man to be 'a mass of electrified clay possessing the power to confine, fetter and deteriorate the omnipresent intelligence of the universe'. Over 200 years later, 'electrified clay' remains a fair description of the human brain.

Although we may scoff at the idea that electricity could animate a lifeless creature and know that a lightning strike is often lethal, even today we are not immune from the idea that electricity is the spark of life. A late-night arts programme on British television (*The South Bank Show*) is introduced by a modified version of Michelangelo's famous painting of God creating Adam, in which an electric spark leaps from God's outstretched finger. Nor is the idea entirely fanciful for, like almost all organisms, humans are electrical machines. As this chapter shows, the development of our knowledge of 'the body electric' is intimately entwined with our understanding of electricity itself.

The Dawn of Understanding

On a dry wintry day you may receive a sharp electric shock when you open the car door or grab a metal doorknob, and find that sparks fly and crackle when you pull off a nylon shirt. Petticoats that cling to your legs, clothes fresh from the tumble dryer that stick together, hair that stands on end when you remove your hat, an electric jolt when you kiss someone, the faint battle-rattle of electric sparks, like 'tiny ghosts shooting', as you comb your hair – all these phenomena happen because static electricity builds up on our bodies. In humid atmospheres the charge dissipates quickly, but under dry conditions it can build up to thousands of volts. Close proximity to metal, however, or even another person, will cause it to discharge. Direct contact is unnecessary, as the electricity will jump the gap, generating a spark in the process. The 'electricity' between two people, that special spark, may be more than just lovers' talk.

The science of static electricity starts with the ancient Greeks' fascination with amber. It is from their word for amber, *electrum*, which derives from *elector*, meaning 'the shining one', that we get the word electron, and hence electricity. Because it was usually found washed up on the seashore, amber's origin was always considered mysterious. The historian Demostratus supposed it the crystallized urine of lynxes. Ovid tells a different story. He relates how Phaethon drove Apollo's chariot (the Sun) too close to the Earth and was struck down by Zeus to prevent a catastrophe. His grief-stricken sisters were transformed into poplar trees and shed golden tears of amber that fell into the River Eridanus in which Phaethon drowned.

Of course, we now know that amber is the petrified resin of extinct pine trees and are familiar with it as jewellery, or as the medium in which prehistoric insects have been encapsulated and perfectly preserved. But amber has another interesting and curious property. When rubbed with wool it generates static electricity, causing it to attract light, dry objects like small bits of tissue paper, feathers, specks of wheat chaff, and even your hair; this may be why Syrian women, who used decorative amber weights on the end of their spindles when spinning wool, called it the 'clutcher'. Thales of Miletus is credited with being the first to note the attractive properties of amber, in the fifth century BC, although it is hard to be certain, as his findings were only passed down orally until later philosophers, such as Theophrastus, wrote them down.

Amber generates a static charge because it attracts electrons from the atoms of the wool, becoming negatively charged in the process and leaving the wool positively charged. The charge is transferred by close contact between the amber and wool – the friction produced by rubbing is not involved, it is simply that rubbing greatly increases the area of contact between the two surfaces. Because opposite charges attract, any material that is naturally positively charged will leap towards the negatively charged amber. Conversely, as similar charges oppose one another, charging up your hair will cause each strand to repel its neighbours as much as possible, producing flyaway hair that stands on end like that of Shock-headed Peter in the German children's picture book. Parenthetically, there is nothing static about 'static' electricity. The term refers only to the fact that the positive and negative electric charges are physically separated. As soon as a positively charged material comes close enough to a negatively charged one, current will flow from one to the other – as visibly demonstrated by the leap of an electric spark.

It was William Gilbert, physician to Queen Elizabeth I, who first invented a sensitive instrument for measuring static electricity (an early electroscope). He used it to compile a long list of materials that could be electrified by rubbing. He also distinguished the attractive power of amber from that of magnets, arguing that two different phenomena are involved. Gilbert was a true scientist, for he advocated that you should not believe what you read, but instead try the experiment for yourself. He wrote, 'Our own age has produced many books about hidden, abstruse, and occult causes and wonders, in all of which amber and jet are set forth as enticing chaff; but they treat the subject in words alone, without finding any reasons or proofs from experiments, their very statements obscuring the thing in a greater fog'. Hence, he concluded, 'all their philosophy bears no fruit'. His words were prescient – present-day scientists make similar complaints about the advocates of astrology and alternative medicine.

Great Balls of Fire

The first machine capable of generating static electricity was invented by the German Otto von Guericke around 1663. It consisted of a giant ball of brimstone, about the size of a child's head, with a wooden rod through its centre. The rod rested in a cradle, enabling the ball to be rotated on its axis by cranking the handle. When a dry hand or a pad of material was held against the whirling ball, static electricity was generated. It is unlikely that von Guericke appreciated that his machine produced electricity, in the modern sense of the word, but he did observe that the globe attracted feathers and other light material, and that once they had first touched the ball, the feathers were repelled and could be chased around the room by lifting the ball from its rod. Careful manipulation even allowed him to balance the feather on another object, such as a colleague's nose.

Frontispiece to *Novi profectus in historia electricitatis, post obitum auctoris*, by Christian August Hausen (1743), depicting the 'flying boy' experiment of Stephen Gray. Von Guericke's ball can be seen on the right. The small boy on the left appears to be standing on an insulating drum, so he will not feel a shock when he touches the flying boy. However, when the gentleman does so, sparks will fly as the current leaps between their fingers and flows through his body to the ground.

One of the most famous uses of von Guericke's machine was the 'flying boy' experiment carried out by Stephen Gray in 1730, for which he was awarded the first Copley medal of the Royal Society of London. The child was suspended in mid-air by insulating cords of silk and then charged up by holding his feet against a rotating sulphur ball. Tissue paper, chaff and other light objects were attracted to his hands, and sparks flew from them when he was discharged.

Large balls of sulphur are not easy to come by, so later electrostatic generators incorporated a circular plate (or spherical globe) of glass that rotated against a fixed cloth; one that was 50 inches in diameter was made for the Emperor Napoleon. The modern equivalent is the Van de Graaf generator, which can produce millions of volts and is well known for its use in spectacular 'hair-raising' demonstrations.

A Jarring Shock

There was no way to store static electricity until the invention of the Leyden jar in October 1745 by a German cleric, Ewald Jürgen von Kleist. Just a few months later the Dutch scientist Pieter van Musschenbroek reported a similar, independent, discovery to the Paris Academy of Sciences. His letter was translated by Jean-Antoine Nollet, the Abbot of the Grand Convent of the Carthusians in Paris, who named the device the Leyden jar, after Leiden, the Netherlands city in which Musschenbroek worked.

The Leyden jar resembles an empty glass jam jar coated inside and out with a thin layer of metal foil that extends about two-thirds of the way up its sides. A brass rod is inserted through an insulating cork stopper into the neck of the jar and connected to the inner metal foil by a chain. If the outer layer of foil is grounded, the inside can be charged up by connecting the rod to a static electricity generator. This happens because the glass wall of the jar acts as an insulator and prevents the charge from flowing to the outer layer of foil, so that a very high charge difference can be built up between the two metal layers. The device is discharged by connecting the inner and outer layers of foil, either with two wires – which generates impressive electric sparks as the wires approach one another – or, more inadvisably, with one's hands.

The charge stored in a Leyden jar can be considerable – and extremely dangerous – as van Musschenbroek discovered. He wrote, 'my right hand was struck with such force that my whole body quivered just like someone hit by lightning [. . .] the arm and the entire body are affected so terribly I can't describe it. I thought I was done for.' He also said that he would not repeat the experiment if offered the whole kingdom of France and cautioned others not to try it. But of course they did, with predictable effects. Some had convulsions or were temporarily paralysed. One German professor who got a severe shock and a bloody nose refused to test it on himself again and instead next tried it on his wife!

The effects were clearly well known to Jules Verne, who described a fantastical device in *Twenty Thousand Leagues under the Sea*. In the novel, Captain Nemo explains to Monsieur Arronax that his underwater rifle fires glass capsules, which are 'exactly like miniature Leyden jars, into which the electricity has been forced at a very high voltage. They discharge at the slightest impact, and however powerful the animal, it falls down dead'. His account contains some artistic licence, but shows how dangerous the Leyden jar was considered to be.

The severity of the shock from a Leyden jar surprised the experimenters because it was much stronger than that of a single spark produced by an electrostatic generator. This was because the jar could accumulate and store the charge flowing in many sparks, which would then be released all at once. Initially, it was believed that electricity was a fluid, so it seemed natural to use bottles and jars to store it in, but it was later appreciated that this was not the case and today the Leyden jar has been replaced by the capacitor. This operates on the same principle. It consists of two parallel metal plates separated by a thin layer of a non-conductive material such as mica, glass or air. The amount of charge a capacitor can store is determined by the area of the plates and the distance between them, and it can be considerable. The first atom smasher, built in the 1930s at Cambridge University by John Cockcroft and Ernest Walton, used banks of capacitors to generate and store up to almost a million volts.

Nine Lords a-Leaping

Another early demonstration of the effects of electricity on the human body was that conducted by the Abbé Nollet. In 1746, he ordered 200 of his monks to form up in a large circle almost a mile in circumference, holding long iron rods between their hands. Once they were all in position, the Abbé surreptitiously connected the two ends of the circle to a Leyden jar. The results were spectacular because the discharge of the jar sent a shock wave through the circle that caused all the monks to jump in turn, thereby demonstrating that electricity travels extremely fast. The French Academician Le Monnier wrote, 'it is singular to see the multitude of different gestures and to hear the instantaneous exclamations of those surprised by the shock'. Hearing of the performance, King Louis XV demanded a rerun at Versailles and a company of 180 soldiers holding hands leapt simultaneously. Adam Walker, a popular British electrical performer in the late eighteenth century, went even further, boasting, 'I have electrified two regiments of soldiers, consisting of eighteen hundred men.'

Such experiments created a sensation. Public demonstrations of electrical phenomena rapidly became the rage, and itinerant lecturers roamed the country. Their aim was more spectacle than science, and performances were generally advertised for their entertainment qualities as much as their educational content. One of the most famous presenters was Benjamin Martin, a consummate entertainer who introduced a season of lectures at Bath in 1746 in which he used a Leyden jar to produce luminous discharges, and 'wonderful Streams of Purple Fire', which looked both beautiful and exotic in a darkened room. Like the Abbé Nollet, he also excited his audience by getting them to join hands and then applying an electric shock, which was not 'so violent and dangerous as they have been represented, tho' they are nearly as great as any Person (especially the *Men*) care to endure'. One letter of the time commented that these public spectacles were 'the universal topic of discourse. The fine ladies forget their cards and scandal to talk of the effects of electricity'.

On other occasions members of the public were invited to be charged up with static electricity and then ignite brandy or ether with sparks from their fingertips. Ladies donned glass slippers to insulate them from the ground and were electrified so that when their gentlemen friends approached with puckered lips outstretched, sparks flew between their lips. The electrified Venus, as she was known, gave stinging kisses. Electric toys abounded. Hidden words were magically revealed using 'fulminating boards' in which sparks jumped between small gaps in a conducting track, paper dancers were animated by the attractive and repulsive forces of static electricity, thunder houses were used to demonstrate the effects of lightning on buildings. Even more dramatic were the pistols and toy cannons that were fired using the heat of an electric spark.

Many of these early demonstrations – and their protagonists – were viewed with suspicion because it was believed that electricity was a manifestation of the force of life and to tamper with it was blasphemy. For others it was a form of fire, which is why Mary Shelley subtitled her book *The Modern Prometheus*, after Prometheus, the mythical Greek who stole fire from the gods to give to mortals.[2] At best, electricity was considered simply a novelty, an entertaining curiosity of no practical value. At which point Benjamin Franklin entered the scene and changed that view forever. In his hands, electricity left the salon and become the province of science.

Snatching Lightning from the Sky[3]

Franklin is widely believed to have been the first to show that lightning is a form of electricity. His most famous experiment was carried out in June 1752, when he flew a kite as a thunderstorm approached to prove that lightning is a stream of electrified air. He connected a short, stiff, pointed wire to the top of the kite, tied a metal key to the end of the kite string and attached the key to a silk ribbon, so insulating it from the ground. Whenever a thundercloud passed overhead, Franklin observed that the loose fibres of the hemp string would stand on end, suggesting that the twine became electrified. He even noted that a stream of sparks would leap from the key to his fingers, and that it was possible to charge a Leyden jar by touching it to the key. Franklin was fortunate not to have been struck by lightning, as this was a very dangerous experiment.

But Franklin was not the first to demonstrate that lightning is an electrical discharge. That accolade goes to a Frenchman, Thomas-François Dalibard. In May of the same year, Dalibard erected an inch-thick, 40-foot-high iron pole, carefully insulating it from the ground by standing it on a plank balanced on three wine bottles and securing it with silken ropes. Sparks could be drawn from the rod with a Leyden jar when lightning was in the area. As Dalibard acknowledged, his experiment was inspired by Franklin's paper describing his 'Experiments and Observations' on electricity, in which the American conjectured that such a pointed rod should draw lightning from the cloud and advised on how harm to the experimenter might be avoided. Dalibard's demonstration created a sensation throughout Europe and was rapidly repeated by many others. Alas, not all were as careful or as lucky as Dalibard. The Swedish scientist Georg Wilhelm Richman was electrocuted a year later while experimenting with lightning conductors; his death is commemorated in a rather flowery poem by Erasmus Darwin (grandfather of the more famous Charles), whose narrator –

> eyed with fond amaze
> The silver streams, and watch'd the sapphire blaze;
> Then burst the steel, the dart electric sped
> And the bold sage lay number'd with the dead!

The Franklin Memorial in Philadelphia is inscribed with some of the statesman-scientist's words of wisdom: 'If you would not be forgotten as soon as you are dead and rotten, either write things worth reading or do things worth the writing.' Franklin, of course, did both. One of his lasting legacies is the lightning conductor. Being aware that lightning was simply a form of electricity, and knowing that lightning strikes the tallest objects, he advised fixing on the 'highest Parts of those Edifices upright Rods of Iron, made as sharp as a Needle and gilt to prevent Rusting, and from the Foot of those Rods a Wire down the outside of the Building into the Ground'. These pointed rods, he surmised, would conduct the strike safely to the ground so the building would not be damaged – or as he more poetically phrased it, 'secure us from that most sudden and terrible Mischief!'

Initially support for Franklin's idea was not universal. Some objected that it would attract lightning to the house, thereby increasing the danger. Others considered it presumptuous as it interfered with the will of God; in Franklin's time many people believed lightning was God's punishment upon the sinful. Franklin countered that lightning was 'no more supernatural than the Rain, Hail or Sunshine of Heaven, against the Inconveniences of which we guard by Roofs and Shades without Scruple'. His argument, and the manifest value of his invention, soon led to the installation of lightning conductors on most gunpowder stores, and even cathedrals.

But in England there were problems. An acrimonious debate broke out between those who supported Franklin's idea of a pointed tip to a lightning conductor and those who preferred a round knob, on the grounds that a sharpened point was dangerously effective and attracted the lightning to it. The latter idea was championed by Benjamin Wilson. He campaigned vigorously against Franklin and he had powerful friends. Matters came to a head in 1777, when the gunpowder magazine administered by the Ordnance Board at Purfleet on the Thames was struck by lightning, and a few bricks were dislodged. The pointed rods previously installed on the advice of Franklin and his colleagues had seemingly not protected the building. Wilson took full advantage of the disaster, producing a spectacular electrical extravaganza at the Pantheon designed to prove that high spikes were dangerous and low blunted knobs were to be preferred. It was performed in the presence of King George III and many prominent ministers, who were impressed by his arguments. The fact that this took place at the time of the American War of Independence added a further charge to the issue. What had begun as a scientific spat quickly escalated into a major feud between the British knob and the American spike factions, with Wilson proclaiming that it was Britain's patriotic duty to dismiss the invention of the enemy. Franklin's friends countered with equally damaging political attacks. The Royal Society waded in, carried out a series of experiments and concluded that Franklin was correct. King George, however, sided with Wilson, ordering pointed spikes to be removed from all royal palaces and Ordnance buildings and demanding the Society reverse its conclusions. But John Pringle, the President of the Society, declined to do so, memorably stating that 'duty as well as inclination would always induce him to execute his Majesty's wishes to the utmost of his power; but "Sire [. . .] I cannot reverse the laws and operations of nature"'. The king promptly suggested he had better resign. Shortly after, a witty friend of Franklin's lampooned the king in the following epigram:

> While you, great George, for knowledge hunt,
> And sharp conductors change for blunt,
> The nation's out of joint:
> Franklin a wiser course pursues,
> And all your thunder useless views
> By keeping to the point.

It was not all plain sailing in France either. Monsieur de Vissery of Arras was ordered to remove a lightning rod he had attached to the chimney of his house. He appealed. By the time the case reached the provincial court of last appeal in 1783, after three years of argument, the case had become the talk of Paris and a political lightning rod. An obscure young lawyer called Maximilien Robespierre made his name by defending science against superstition and winning the case, arguing that while theory required experts to interpret it, the facts did not.

Ten years later, the National Convention, led by Robespierre, used a similar argument to get rid of government experts and all national academies and literary societies. Robespierre is best known for instituting the Reign of Terror during which many French aristocrats were guillotined. It is possible that without his successful defence of Monsieur de Vissery and his lightning conductor, Robespierre might not have moved to Paris and the course of French history might have been very different.

Today, almost all tall buildings sport lightning rods similar to those advocated by Franklin, that lead the electric current safely to the ground and spare the building. Large structures may have several of them. St Paul's Cathedral in London, for example, has them spaced at regular intervals around the roof. And they are essential: the Empire State Building is regularly hit during a lightning storm, demonstrating that the axiom 'lightning never strikes twice in the same place' is a dangerous fallacy.

Franklin advised that it was not wise to shelter under an isolated tree in a storm, as it was likely to attract a lightning strike. He also noted that wet clothing provides a low resistance path to ground (outside the body), so that the current flash preferentially runs over the surface of the body rather than through it, and he concluded that this was why a 'wet Rat can not be kill'd by the exploding Electric Bottle when a dry Rat may'. His idea may explain why one young man hit by a lightning bolt survived unscathed, for he was wearing an oilskin (rain slicker) that was soaking wet from a torrential rainstorm. His father witnessed the lightning strike from the safety of his pick-up truck and rushed his son to hospital; but he was discharged an hour later with no ill effects. Most people are not so lucky, and lightning strikes kill and maim hundreds of people every year.

Bolts from the Blue

Lightning is bred in cumulonimbus, those towering anvil-shaped clouds with billowing sides and flat bottoms that form when warm moist air rises to a height at which it is cold enough to freeze water. In such thunderclouds, ice particles and water droplets are continually colliding as air movements swirl them about. Tiny ice crystals become positively charged and are tossed to the top of the cloud, whereas bulkier chunks of ice and slush, the size of small hailstones, become negatively charged and sink to the bottom. This creates a charge separation, with upper layers of cloud having a positive charge and the lower ones a negative one. The voltage difference between the negatively charged lower layers of the cloud and the ground can reach as much as 100 million volts. At some point this difference is so great that it exceeds the insulating capacity of the air, and the current arcs to ground in a lightning flash. It lasts only a fraction of a second. There is also a rare form of lightning in which the bolt issues from the top of the cloud. Such 'positive lightning' is highly dangerous, as it can strike ground many miles from the cloud, without warning, on a sunny day – a veritable bolt from the blue.

A lightning bolt can reach speeds of 60,000 metres a second and temperatures of 30,000°C, five times hotter than the surface of the Sun. It averages three miles long, but is only about a centimetre wide. Each flash is actually made up of several individual discharges that occur too fast for the eye to distinguish them fully, which explains why lightning appears to flicker. A single strike unleashes as much energy as a ton of TNT and the intense heat induces an explosive expansion of the air at speeds that break the sound barrier, which is heard as a thunderclap. Although thunder and lightning are generated simultaneously, light travels much faster than sound; 186,000 miles a second as compared to a mere 0.2 miles a second. Thus you see the flash first and hear the thunder some time later, depending how far away the storm is.

Thunderstruck

If you are unfortunate enough to be hit by lightning, some of the current will flow over the surface of your body and some through your body, with the relative proportions depending on which path offers the least resistance. The former is less dangerous and it is likely that people who survive a strike mainly experience such a 'flashover'. If you and your clothes are soaked by rain the water turns to steam, which can blow off your garments and burn your skin. Current that flows through the body can cause serious internal damage. Many people hit by lightning suffer cardiac arrest and require immediate cardiopulmonary resuscitation to avoid brain damage (people hit by lightning do not remain charged and can be safely touched). The respiratory centres in the brain may also be affected so that the person ceases to breathe. There are reports of people who were unable to breathe spontaneously for up to twenty minutes after a strike, despite cardiac function returning, so it is essential to continue artificial ventilation of victims who are apparently dead, as resuscitation is often successful. Neurological symptoms such as loss of consciousness, confusion, memory loss and partial paralysis, especially of the lower limbs, are very common. Other effects include hearing loss, blindness, sleep disorders and severe burns. The electric current can also stimulate muscle contraction, which is why people appear to jump or be blown across the room when struck. As all your muscles contract at once, they catapult you into the air.

The Frog's Dancing Master

The dramatic effects of an electrical discharge, exemplified by electrostatic generators and lightning, led many eighteenth-century experimenters to seek to understand its physiological effects. Among them was the great Italian scientist Luigi Galvani, who was the first to discover 'animal electricity'. Although Galvani had originally intended to enter the Church, his parents persuaded him to study medicine and by 1762 he was established as a professor of anatomy in his hometown of Bologna. Like many scientists of his day, he was interested in static electricity, and as early as 1780 he began to study its effects on muscle contraction. He worked with a small team that included his wife Lucia and his nephews Camillo and Aldini in a laboratory set up in his home.

Plate 1 of Galvani's *Commentarius* showing several sets of dissected frogs' legs. An electrostatic generator sits on the left of the table and a Leyden jar on the right. The small lace-cuffed hands pointing to the instruments, reminiscent of those used in the Monty Python films, were a common way to mark a point in the Renaissance.

On 26 January 1781, Galvani's journal records, he serendipitously noticed that when his assistant touched the nerve that supplies the leg of a recently dead frog with a metal implement all the muscles in the legs twitched violently. However, this only happened at the precise moment that a spark was generated by the discharge of an electrical machine. Galvani repeated the experiment many times and in many different ways, but always got the same result. Consequently, he hypothesized that the electric spark was stimulating the muscle to contract. This led him to wonder if lightning could also cause the frog's muscles to shorten. He tested this idea with the help of his nephew Camillo by attaching the nerve that supplies the frog's leg muscle to a long wire that was connected to a metal spike that he placed at the top of his house, pointing towards the sky. As he had predicted, he found that the frog's legs twitched wildly when the lightning flashed during an electrical storm.

Being a careful scientist, Galvani repeated the experiment on a calm day, as a control. This time he suspended the frog's legs from the iron railings of his balcony by brass hooks that pierced the spinal cord. At first, nothing happened. Getting impatient, Galvani began to fiddle with the frog's legs. To his surprise, he noticed they then began to display frequent spontaneous and irregular movements, none of which depended on the variations of the weather, but which occurred when the hooks were pressed against the railing.

Galvani interpreted this result to indicate that animal cells are not only stimulated by electricity – they can actually produce their own. This electrical (self)-stimulation, he surmised, produced contraction of the muscle. In 1791 he published his discoveries in a pamphlet entitled *De viribus electricitatis in motu musculari commentarius* ('A commentary on the effects of electricity on the motion of muscles') in which he contended that animal electricity was different in kind from that produced by lightning or an electrostatic generator, and he argued that 'the electricity was inherent in the animal itself'. Galvani had a few copies of his article published at his own expense and sent them to his fellow scientists, including his friend and countryman, Alessandro Volta, who was professor of physics at the University of Pavia.

At first his colleagues accepted Galvani's ideas with cautious enthusiasm, repeating his experiments and obtaining the same results. As a consequence, the supply of live frogs diminished precipitously and a year after Galvani's publication Eusebio Valli was complaining to a colleague, 'Sir, I want frogs. You must find them. I will never pardon you, Sir, if you fail to do so. I am without ceremonies, Your most humble servant, Valli'.

Volta's experiments caused him to revise his initial conclusion that Galvani was correct, and argue instead that the muscle twitches Galvani had observed in the absence of extrinsic electrical stimulation were not due to an innate animal electricity. Rather, he deduced (correctly) that they were induced by an electric current flowing between two dissimilar metals – the iron railing of the balcony and the brass hooks which Galvani had attached to the nerve supplying the frog's leg. This initiated a heated controversy between the two scientists about whether the origin of the stimulus was biological or physical.

While Galvani acknowledged Volta's argument, he remained convinced that animal electricity was a real phenomenon. Crucially, he showed that putting the nerve in contact with the muscle was enough to cause a spasm – no metal at all was needed. We now know that this experiment worked because the injured tissue generates an electric current that is able to stimulate the muscle, although Galvani did not realize this. Unfortunately, he published this experiment anonymously, which somewhat diminished the force of his argument.

Power to the People

The fact that contact between the nerve and muscle was sufficient to cause contraction was a triumph for galvanism and a setback for Volta. But he was not defeated and continued to explore the idea that contact between dissimilar metals was involved. Believing that electricity was not of animal origin, he decided to dispense with the frog altogether. He built a stack of alternating silver and zinc discs, separated by cardboard soaked in salt water, and demonstrated that an electric current flowed when the top and bottom of the pile were connected. He had invented the first electric battery. Indeed, he got an electric shock by touching one hand to the top and the other to the bottom of the pile. Volta also drew attention to the striking resemblance of his apparatus to the electric organs of electric eels and rays, fish well known for their ability to give humans a substantial electric shock. The electric organs of these fish consist of stacks of cells separated by conducting fluid, analogous to Volta's pile of silver and zinc discs.

The shock produced by Volta's early battery was feebler than that of a Leyden jar, but it had a singular advantage: it was produced indefinitely and did not need to be charged in advance from an electrostatic generator. Larger currents – and thus bigger shocks – could be produced by increasing the height of the pile of discs. Volta described his invention in a letter to the Royal Society of London in 1800 entitled 'On the electricity excited by the mere contact of conducting substances of different kinds'. Written in French by an Italian scientist to an English institution, it demonstrates that even in 1800 science was an international activity. Volta later presented one of his own voltaic piles to the Royal Institution in London where it can still be seen today.

Clash of the Titans

The disagreement between Galvani and Volta over the interpretation of the frog experiments is sometimes portrayed as a scientific feud, with Galvani being the loser; and the invention of the battery was seen a triumph of Volta over Galvani, and of physics over biology. But Galvani was not completely wrong, because his idea that animals produce electrical signals in their nerves and muscle fibres turned out to be correct. Unfortunately, the fact that Volta's ideas took precedence probably set back the science of animal electricity for some time.

Although the issue divided the scientific community, and their supporters fought about the matter, the dispute between Volta and Galvani themselves was not acrimonious. Volta wrote of Galvani's work that it contained 'one of the most beautiful and most surprising discoveries' and he is credited with inventing the term 'galvanism'. He even communicated Galvani's findings to the Royal Society of London, writing 'an account of some discoveries made by Mr Galvani, of Bologna; with experiments and observations on them'. Interestingly, in his very first sentence he refers to the subject of his letter as discoveries and researches on 'L'Electricité Animale', although he goes on to conclude that it does not exist.

Perhaps some of the reason that Galvani's ideas fell into abeyance is that he was a less effective champion than his rival. Galvani was a rather retiring individual. He only published his work in 1791, at least ten years after his initial experiments, and he did so (in Latin) in the transactions of the Bolognese Istituto delle Scienze, which were not widely available. He compounded the problem by being averse to travel, a poor correspondent, failing to publish some of his experiments altogether and generally communicating his findings only to his immediate circle. Political problems also hindered Galvani's work. In 1794 Napoleon conquered Bologna and two years later Galvani was forced to resign his professorship because he refused to take the loyal oath to the French Cisalpine Republic demanded by the university as it contravened his political and religious principles. He fled to the home of his brother Giacomo, where he broke down in despair. His friends obtained an exemption from the oath for him, on account of his scientific accomplishments, but tragically he died before it could be implemented. He was only sixty-one.

Volta had a very different temperament and life. He was a charismatic and dynamic speaker and a prolific (occasionally arrogant) author who published in several languages, promoted his work widely and willingly accepted the new regime. Volta became very well known and was fêted throughout Europe. In 1801, he was invited to Paris, where he was presented with a gold medal and gave three lecture demonstrations on his findings, all of which were attended by Napoleon. Volta collected many other prizes and distinctions, being appointed to the Legion of Honour by Napoleon in 1805 and later being made an Italian senator and subsequently a count. The unit of electrical potential was also named the 'volt' in his honour. Far more politically astute than Galvani, Volta continued to find favour even after Napoleon fell and power shifted to Austria.

The 'Mad' Scientists

Galvani's experiments generated considerable excitement. All across Europe, scientists and laymen alike tried to reproduce his findings, not only on recently dead frogs but also on other dead animals.

Galvani's flamboyant nephew, Giovanni Aldini, brought electricity to public attention in a particularly sensational way. An extraordinary synthesis of scientist and showman, his (in)famous public demonstrations may have inspired Mary Shelley's novel *Frankenstein*, for not content with merely applying electric shocks to frogs' legs, Aldini used the bodies of recently executed criminals. Ironically, given his fierce antagonism to Volta, he had to rely on a voltaic pile to generate the necessary shock.

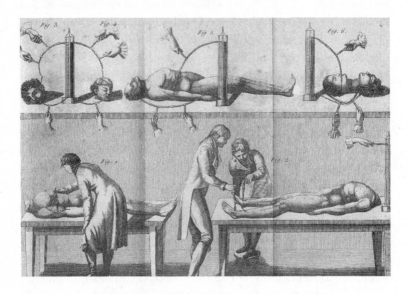

Illustration from Aldini's treatise on his experiments on the decapitated cadavers of criminals, *Essai théorique et éxperimental sur le galvanisme*. The tall, pencil-like structure is a voltaic pile (a primitive battery), which is used to generate an electric current. The current is applied to the corpse via a curved metal rod attached to an insulating glass handle that is held by the experimenter to prevent him getting a shock.

Aldini's notes record that in 1802, 'The first of the decapitated criminals was transported to the room I had chosen, close to the place where the execution was carried out. The head was first subjected to galvanic action using a stack of a hundred pieces of silver and zinc. A metal wire was placed inside each of the two ears, which had been moistened with salt water. The other end of the wire was connected to either the top or the bottom of the pile. I initially observed strong contractions in all the muscles of the face, which were contorted so irregularly that they imitated the most hideous grimaces. The action of the eyelids was particularly marked, though less striking in the human head than in that of an ox.'

His most notorious demonstration took place in London on 17 January 1803 when he electrified the corpse of the murderer Thomas Forster. Immediately after his execution (by hanging), the body of the malefactor was conveyed to the Royal College of Surgeons where a large audience awaited. Aldini took a pair of conducting rods, each connected at one end to a voltaic pile, and applied the other end of the first one to the corpse's mouth and that of the second to the ear, whereupon 'the jaw began to quiver, the adjoining muscles were horribly contorted, and the left eye actually opened'. When the rods were applied to the dissected thumb muscles they 'induced a forcible effort to clench the hand'. In another experiment, violent convulsions were produced in all the muscles of the arm. The highlight of the demonstration came when rods were applied to the ear and rectum, which 'excited in the muscles contractions much stronger [. . .] so much increased as almost to give an appearance of re-animation'.

But Aldini was not the first to use humans. As early as 1798, Xavier Bichat experimented upon bodies of those guillotined during the French Revolution, within forty minutes of their execution. He had no shortage of subjects. His experiments showed that the heart could be excited by electricity when stimulated by direct contact, and they exerted a macabre fascination on the scientific and literary imagination alike.

Equally grotesque experiments were carried out in 1818 by Dr Andrew Ure, which produced as electrifying a reaction in the audience as in the corpse. As Ure describes in his book *A Dictionary of Chemistry and Mineralogy*, the cadaver was that of an extremely muscular young man who was brought to the anatomy theatre of Glasgow University within ten minutes of being cut down. Incisions were made in the body to enable a battery to be connected, via conducting rods, directly to the nerves. Application of one rod to the spinal cord and the other to the sciatic nerve resulted in violent shudders throughout the body. And, 'On moving the second rod from the hip to the heel, the knee being previously bent, the leg was thrown out with such violence as nearly to overturn one of the assistants, who in vain attempted to prevent its extension.' In a second experiment, the rod was applied to the phrenic nerve in the neck. The success of this was 'truly wonderful. Full, nay, laborious breathing, instantly commenced. The chest heaved, and fell; the belly was protruded and again collapsed, with the relaxing and retiring diaphragm'. Touching the rod to the supraorbital nerve induced the most extraordinary grimaces – 'rage, horror, despair, anguish, and ghastly smiles, united their hideous expression in the murderer's face'. Several of the spectators were forced to leave the room from terror or sickness and one man fainted. But worse was to come for electrical stimulation of the ulnar nerve animated the fingers, which 'moved nimbly like those of a violin performer', and at one point the arm shook and the forefinger extended and seemed to point at the spectators, some of whom thought it had come to life.

Such spectacles did not help the popular view that all doctors were quacks. Small wonder that Lord Byron wrote,

> What varied wonders tempt us as they pass!
> The Cow-pox, Tractors, Galvanism, and Gas
> In turns appear, to make the vulgar stare,
> Till the swoln bubble bursts – and all is air!

The blasphemous nature of the experiments, due to the possibility of 'resurrecting' the dead, also did not go unremarked. Together with Frankenstein's monster, they gave rise to the idea of the scientist as both 'mad' and 'bad', an image that pervades the media even today.

With present-day knowledge, the experiments of Galvani and his colleagues are easily explained. The cells of the body do not die when an animal (or person) itself breathes its last breath, which is why it is possible to transplant organs from one individual to another, and why blood transfusions work. Unless it is blown to smithereens, the death of a multicellular organism is rarely an instantaneous event, but instead a gradual closing down, an extinction by stages. Nerve and muscle cells continue to retain their hold on life for some time after the individual is dead and thus can be 'animated' by application of electricity. Just as an electric shock will stimulate your nerves so that the muscles they innervate will contract, so too with the nerves of a recently dead corpse. Indeed, the experiments of Ure and Aldini provide a dramatic demonstration of which muscles individual nerves innervate. Nevertheless, the sooner after death the experiment can be implemented, the more likely a response will be obtained.

The Age of Wonder

By the end of the eighteenth century, electricity could be generated, stored and conducted along wires for significant distances. Its wondrous effects excited the interest of scientists and galvanized their research. The culture of the Enlightenment, which dictated that scientific advances should be communicated to the non-specialist, led to spectacular public entertainments that sparked interest in the wider community. Indeed the public science lectures given by the director of London's Royal Institution, Michael Faraday, were so popular with the wealthy gentry that Albemarle Street became heavily congested, particularly when carriages were collecting people after a lecture, and the first one-way street in the city had to be introduced.

The use of electricity to treat all kinds of medical complaint was widely advocated, as related in Chapter 12. Lightning rods and early batteries offered other practical applications and heralded the dawn of a new electric age. But not everyone was immediately impressed with the possibilities that electricity afforded. William Gladstone, then Chancellor of the Exchequer, once visited Faraday's laboratory. He stood silent for a moment before the scientist's electrical contraptions and then remarked, 'It is very interesting, Mr Faraday, but what practical worth has it?' Faraday was more than a match for him. He is reputed to have replied, 'Sir, I know not what these machines will be used for, but I am sure that one day you will tax them.'

It was also appreciated that nerve and muscle fibres could be stimulated by electric shocks. Although Galvani's idea of animal electricity was contested, it was not without support, for it had been known since antiquity that electric fishes could produce a severe electric shock. Furthermore in 1797 the young scientist and explorer Alexander von Humboldt established that Galvani's and Volta's ideas were both correct and predicted that every contraction of a muscle would be preceded by an electrical discharge from the nerve supplying it. In this light, the idea that galvanism could animate a lifeless creature, such as Victor Frankenstein's monster, requires only a small stretch of the imagination. Recording the currents associated with the conduction of nerve and muscle impulses and unravelling the underlying mechanisms, however, had to await the development of suitable instruments and a greater understanding of electricity itself.

2

Molecular Pores

The American quarter horse
known as Impressive,
the shivering pig,
a whole herd of goats
in Texas, and some
among you in the front row
with your various flaws
will feel the pang of recognition,
a flutter in the ion channels,
as you watch me fall down.

Jo Shapcott, 'Discourses'

During an oral examination at Oxford University around 1890, a student was asked if he could explain electricity. He replied nervously that he was sure he once knew what it was – but he had forgotten. 'How very unfortunate', remarked the examiner, 'Only two persons have ever known what electricity is: the Author of Nature and yourself. Now one of the two has forgotten.'

Today, we are all very familiar with electricity as it powers our industrial society. Almost everything we use – our transport, lighting and communications devices, even the computer I am writing this on – runs on electricity. But what is less widely recognized is that we too are electrical machines and that electrical currents lie at the heart of life itself. In turn, these currents are due to the activity of ion channels. To appreciate how it has been possible to leap from Galvani's experiments with frogs' legs to our ability to treat diseases of electrical activity like epilepsy – or the neonatal diabetes from which James suffers – it is necessary to understand what ion channels are and how they contribute to the electrical responses in cells.

For more than a century and a half after Galvani, scientists laboured to measure the electrical impulses of our nerves and interpret what they meant. It took even longer to detect the ion channels that are responsible – but their discovery transformed our understanding. The concepts I had struggled with as a young student, and that caused me many sleepless nights (especially around the time of the examinations), suddenly became crystal clear. This chapter therefore jumps straight to the present day and provides a state-of-the-art picture of how ion channels work. But first it is helpful to consider what electricity is and how the electricity in your head differs from that supplied to your home.

The Holy Trinity

Electricity is a form of energy that is based on electric charge, one of the most fundamental properties of subatomic matter. The electric currents that flow through the wires in our houses – and along our nerves – are quantified in terms of three basic units: the amp (A), the volt (V), and the ohm (Ω). They are named after a triumvirate of great eighteenth-century European physicists: the Frenchman André-Marie Ampere, the Italian Alessandro Volta and the German Georg Ohm. Current is measured in amps, resistance to current flow in ohms, and voltage, the force that drives the current flow, in volts.

The flow of electric current through a wire is often explained by analogy with the flow of water through a pipe. In electrical terms, the current corresponds to the rate at which a stream of charged particles moves, with one amp being the equivalent of approximately six million million million ($6x10^{18}$) particles per second.

Resistance is a measure of the ease or difficulty of flow. Narrowing a pipe will restrict the flow of water, whereas increasing the pipe's diameter will increase water flow. In an electric circuit, materials that offer little resistance to current flow, such as metals, are called conductors, whereas those that resist the flow of current (like paper or air) are known as insulators. Grasp the bare wire of an electric fence and you will get an unpleasant shock, but you will feel nothing if you instead take hold of one of the insulated handles that enables you to open the gate in the fence.

The voltage difference between one point and another is analogous to the difference in water pressure that causes water to flow from one region to another. Essentially it is the force that drives the current flow. It is also sometimes called the electrical potential difference (or potential for short). Providing two points are not connected, no water will flow between them, and in an analogous way an electric current will only flow when a circuit is complete. This is why there can be a huge voltage difference between a thundercloud and the ground, but no current flows until the lightning bolt jumps the gap. It also explains why electrons do not move along a wire unless the electric circuit is complete, and thus why your desk lamp does not light up until you flick the on/off switch that links the wires. Just as increasing the water pressure will increase the flow of water, so boosting the voltage increases the current. Increasing the voltage supplied to the lamp, for example, will make it shine brighter.

Ground (or earth) is defined as the lowest point of voltage and, like water, current will always flow to the lowest level. This was appreciated early. In 1785 Joseph-Aignan Sigaud de la Fond was bewildered to find that although he had a highly charged Leyden bottle and a chain of sixty people holding hands, the shock was not felt by more than the first six. Why it stopped at the sixth man was a mystery, and it was argued that there must be something peculiar about him. The favoured hypothesis was that the young man in question was not endowed with 'everything that constitutes the distinctive character of a man' – not, in other words, in full possession of Nature's attributes. Gossips quickly spread the idea around Paris that it was impossible to electrify a eunuch.

The Duc de Châtres, who had a scientific turn of mind, demanded proof and the requisite experiment was carried out using three of the King's musicians, with understandable apprehension on the part of both the experimental subjects and the fully endowed controls. To Sigaud's further bafflement, all three castrati felt the shock. The mystery was only solved when the experiment was repeated many more times and it was noticed that individuals who did not transmit the shock were standing on soggy ground. As wet earth is a better conductor of electricity than the human body, the current preferentially flowed to ground. This is also the reason you get an electric shock when you accidentally touch a live wire: the ground on which you stand is at a lower voltage than the wire in your hand, so that the current flows through you to the ground.

Amps, volts and ohms are bound together in an eternal embrace, as was first appreciated by Ohm. He formulated a famous law, which states that the current (I) is equal to the potential (V) divided by the resistance (R); it is abbreviated mathematically as $I = V/R$. In other words, provided that the resistance remains the same, increasing the voltage will increase the magnitude of the current that flows. Similarly, if the resistance falls, but the voltage remains the same, then the current will increase. And so on. This simple formula, known as Ohm's Law, is the key to understanding how nerves – and electricity – work.

Poles Apart

There is a fundamental difference, however, between the electricity that powers our bodies and that which lights our cities. The electricity supplied to our homes is carried by electrons. These indivisible subatomic particles carry a negative electric charge and because opposite charges attract one another (and similar charges repel) electrons always flow from a region of negative to positive charge. Confusingly, we define current as the direction of flow of positive charges, which means that the current in a wire moves in the opposite direction to that in which the electrons flow!

In contrast, almost all currents in the animal kingdom are carried by ions – electrically charged atoms. There are five main ions that carry currents in our bodies. Four are positively charged – sodium, potassium, calcium and hydrogen (protons) – and one, chloride, is negatively charged. Because they are electrically charged, the movement of ions creates an electric current. In the case of positively charged ions, the current flow is in the same direction as the flow of ions, whereas for negatively charged ions (as for electrons) it is in the opposite direction.

It is also worth noting that currents in electric circuits flow along the length of the wire. In contrast, the ion currents responsible for nerve impulses flow across the membranes that envelop our cells, into or out of the cell. Thus although electrical impulses travel along the length of our nerve and muscle fibres, the ion currents that generate them flow at right angles to the direction of travel.

Another difference between the electrical signals in our heads and in our homes is their speed of transmission. An electrical signal in a wire travels at almost the speed of light, which is 186 thousand miles per second. It's easy to see this, for when you flick the switch a light comes on at once, and telephones and the Internet provide almost instantaneous communication around the globe. By comparison, the fastest nerve impulses are pitifully slow, crawling along at a mere 0.07 miles a second (120 metres a second). Even the brightest of us cannot think at the speed of light.

As well as being slower, the electrical impulses we generate are also much smaller. Your electric kettle needs three amps of current to run, but the currents that tell your heart when to beat are only a few millionths of an amp. Finally, while energy is needed in both cases, the power – the battery if you will – that drives the current is produced in quite different ways, as explained later.

These differences between animal electricity and that which supplies our homes are simple to state, but took many years to understand. Although the fundamental properties of electricity were understood by the beginning of the nineteenth century, it is only in the last sixty years or so that we have begun to understand the origin of bioelectricity and only in the last fifteen years that we have had a glimpse of what the molecules (the ion channels) responsible for the electrical activity of our nerve and muscle cells actually look like.

The Building Blocks of Life

We are no more than a collection of cells, millions and millions of them – as many as the stars in the galaxy. They come in many different varieties, like muscle cells and brain cells and blood cells, and in multiple shapes and sizes, but they are all the same fundamental entity. Robert Hooke discovered them in 1665 when he was examining a small section of cork under his microscope. He named them cells because they reminded him of the tiny chambers monks lived in, but you'll get a better idea of what they look like if you imagine the cells of a honeycomb on a much smaller scale.

Cells are teeming with molecules carrying out all sorts of complicated reactions, labouring away at making proteins, replicating DNA and generating energy. However, for understanding the electrical properties of cells we only need to consider the events at the cell surface, as it is here that voltage differences arise and nerve impulses are transmitted.

Each of our cells is surrounded by a membrane that encloses its contents and serves as a barrier to the world outside, rather like the skin of a soap bubble. This membrane is made up of fats (technically known as lipids), which means that it is impermeable to most water-soluble substances. It arises from the simple fact that fats and water do not mix. As anyone who has made a vinaigrette salad dressing knows, over time the ingredients separate into a lower layer of vinegar with the lighter oil floating on top. The phospholipid molecules that make up the cell membrane have water-loving phosphate heads and lipid tails that prefer to avoid water, and they organize themselves into a double-layered membrane in which the water-shy lipid tails are sandwiched inside the bilayer between two layers of phosphate head-groups. Don't think, though, that membrane lipids are as hard as butter – they are more the consistency of machine oil, so that the proteins that sit in them tend to float about and must be anchored to the cell's cytoskeleton to keep them in their correct places.

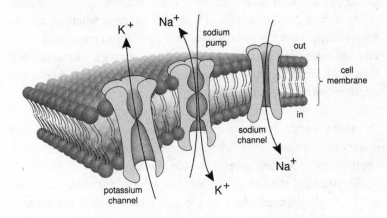

Schematic view of the cell membrane, showing the two layers of lipid molecules, and membrane proteins, such as ion channels and pumps, embedded in it. K^+ is the scientific abbreviation for the potassium ion and Na^+ that for the sodium ion.

The solutions inside our cells, and those of all other organisms on Earth, are high in potassium ions and low in sodium ions. In contrast, blood and the extracellular fluids that bathe our cells are low in potassium but high in sodium ions. These ionic differences are exploited to generate the electrical impulses in our nerve and muscle cells for, like water trapped behind a hydroelectric dam, they are an effective way of storing potential energy. Open the floodgates and that energy is instantly released as the ions redistribute themselves to try and establish equal concentrations on either side of the membrane. It is these ion movements that give rise to our nerve and muscle impulses.

The transmembrane sodium and potassium gradients are maintained by a minute molecular motor, known as the sodium pump, that spans the cell membrane. This protein pumps out excess sodium ions that leak into the cell and exchanges them for potassium ions. If the pump fails, the ion concentration gradients gradually run down and when they have collapsed completely no electrical impulses can be generated, in the same way that a flat battery cannot start your car. Consequently, your sense organs, nerves, muscles – indeed all your cells – simply grind to a halt. This is what happens when we die. As we no longer have the energy to power the sodium pump and maintain the ion differences across our cell membranes, our cells soon cease to function. And while externally applied electric shocks can interfere with the electrical impulses in our nerve and muscle cells, they cannot restore the ion concentration gradients across our cell membranes once they have collapsed. This, then, is why we cannot reanimate a corpse with electricity, and why the spark of life is different from the electricity supplied to our homes.

Maintaining the ion gradients is expensive, for electricity does not come cheap, even when we produce it ourselves. It is extraordinary to think that about a third of the oxygen we breathe and half of the food we eat is used to maintain the ion concentration gradients across our cell membranes. The brain alone uses about 10 per cent of the oxygen you breathe to drive the sodium pump and keep your nerve cell batteries charged. Perhaps surprisingly, it seems that merely thinking is energetically expensive.

The Precious Bodily Fluids

How our cells come to be filled with potassium ions is something of a puzzle. The simplest explanation is that the first cells evolved in a solution high in potassium. Left to themselves, lipids spontaneously organize into liposomes, tiny fluid-filled spheres enclosed by a single skin of lipids. Such lipid films may have been the origin of the first membranes and the liposomes they gave rise to may have formed the precursors to real cells. Over three and a half billion years ago, we may imagine, liposomes engulfed self-replicating molecules such as RNA or DNA[1] and so gave rise to the very first cells.

The fluid enclosed within these first primitive cells would of necessity be the same as that which surrounded them. Thus the high internal potassium concentration characteristic of all cells – from the simplest bacterium to the most complex organism – may reflect the composition of the ancestral soup. This leaves a mystery. Where were those ancient waters rich in potassium? Currently, one popular view is that life evolved in the black smokers of the ocean floor, the hydrothermal vents that belch out superheated water rich in minerals. From a physiologist's point of view, however, this seems rather unlikely, for the Precambrian seas were high in sodium, like those of today. Thus, I side with Charles Darwin, who suggested life evolved in a 'warm little pond'; aeons ago, shallow puddles in which organic molecules could be concentrated, and into which potassium ions leached from the surrounding rocks or clays, may have been the birthplace of the first cells.

At some point in the far distant past, single cells discovered that living together gave them a selective advantage and the first multicellular organisms were born. Because the extracellular solution that bathes our cells is high in sodium, it is likely that such early multicellular organisms evolved in the sea, which is largely a solution of sodium chloride (common salt). It is a fascinating idea that the solutions inside our cells, and those that make up our extracellular fluids, provide a fingerprint to our past history and help chart where life first evolved.

Border Control

The presence of a cell membrane brought numerous advantages. Molecules no longer diffused away from each other at random, but could be retained in close proximity within the cell and, more importantly, interact with one another. Cells could become specialized for different functions – evolving into muscle, liver and nerve cells, to name but a few. Like the walls of a mediaeval city, the membrane also protected the cell from toxins in its immediate environment and restricted substances from entering and leaving, because the lipids of which it is composed are impermeable to most substances. As a consequence, tightly guarded gates that enabled vital nutrients and waste products to enter and leave the cell became a necessity.

These gates are highly specialized transport proteins. They come in many different varieties, but some of the most important are the ion channels. As Primo Levi once said, 'everyone knows what a channel is: it forces water to flow from a source to an outlet between two basically insuperable banks.' However, the term lends itself equally well to describing other types of conduits, including those which facilitate the flow of ions across the cell membrane. In essence, an ion channel is no more than a tiny protein pore. It has a central hole through which the ions move, and one or more gates that can be opened and closed as required to regulate ion movements. When the gate is open, ions such as sodium and potassium swarm through the pore, into or out of the cell, at a rate of over a million ions a second. Conversely, when the gate is closed, ion flux is prevented.

The very largest ion channels are simply giant holes, so big that many ions can go through at a time, and both negatively charged ions (anions) and positively charged ions (cations) can permeate, as well as quite large molecules. This type of channel is rather uncommon and it is easy to see why – all those ion concentration gradients that the cell sets up and protects so carefully would immediately be dissipated if such a channel were to open, causing the cell to die. Indeed, some bacterial toxins kill cells in exactly this way. Most channels, however, are choosy about the ions they allow to pass through their pores. Although some permit access to all cations (and others to all anions), the majority are far more discriminatory. A potassium channel, for example, will only let potassium ions through and excludes sodium and calcium ions, whereas a sodium channel allows sodium to permeate, but not potassium or calcium. As must by now be obvious, channels are generally called after the ion they most favour.

An Electrochemical Battle for Potassium

Under resting conditions all cells have a voltage difference across their membrane, the inside of the cell usually being between 60 and 90 millivolts more negative than the outside. This resting potential arises because of a tug of war between the concentration and electrical gradients across the cell membrane that the potassium ion experiences.

At rest, many potassium channels are open in the cell membrane. As potassium ions are high inside the cell but low outside, they rush out of the cell down their concentration gradient and, because potassium ions carry a positive charge, their exodus leads to a loss of positive charge – or, to put it another way, the inside of the cell becomes gradually more negative. At some point, the exit of potassium ions is impeded by the increasing negative charge within the cell, which exerts an attractive force on the potassium ion that counteracts further movement. The membrane potential at which the chemical force driving the potassium ions out of the cell and the electrical force holding them back exactly balance one another is known as the equilibrium potential.

If the membrane were only permeable to potassium ions, the resting membrane potential would be exactly the same as the potassium equilibrium potential. However, the real world is not so simple and in most cells a few other types of ion channel are open that allow positively charged ions to sneak into the cell, pushing the resting potential to a more positive level.

The importance of the resting potential is that it acts like a tiny battery in which electric charge (in the form of ion gradients) is separated by the insulating properties of the lipid membrane. This stored energy is used to power the electrical impulses of our nerve and muscle fibres.

Ions take the path of least resistance and move down their concentration gradient from an area of high concentration to one of low concentration. The number of sodium ions is much higher outside the cell than inside, so that sodium ions flood into the cell when the sodium channel gates open. Conversely, as there are many more potassium ions inside than out, potassium ions tend to leave the cell when the potassium channels open. Because ions are charged, their flow produces an electric current. It is such currents, carried by ions surging through ion channels, that underlie all our nerve and muscle impulses, and that regulate the beating of our hearts, the movement of our muscles and the electrical signals in our brains that give rise to our thoughts. This, in essence, is how the energy stored in the concentration gradients is used to power the electrical impulses of our nerve and muscle fibres.

Suck it and See

Given the importance of ion channels, it may seem surprising that their very existence was not dreamt of until the middle of the last century and that even by the early 1970s the idea that ions crossed the membrane via specialized protein pores was still a matter of speculation. To demonstrate their existence directly it was necessary to measure the current that flows through a single channel when it opens. This was far from easy, because the current is extremely tiny and can only be measured with highly specialized electronic equipment. If you consider that the currents flowing through a single ion channel when it is open are about a million millionth of the current needed to power your kettle – a few picoamps only – you will get some idea of just how infinitesimally small they are.

The problem was solved using an elegant technique invented by two German scientists, Erwin Neher and Bert Sakmann, which won them a Nobel Prize. Truly innovative science often arises at the interface of different disciplines and their prize-winning combination of talent illustrates this perfectly. Neher was trained as a physicist and Sakmann in medicine, so they brought complementary skills to the problem; together they provided the breadth of vision to see where this new technology could take them and the attention to detail needed to perfect the method. As their colleague David Colquhoun once said, they are 'scientist's scientists' – modest, unassuming, courageous – and inspirational.

Neher and Sakmann reasoned that if ion channels actually existed, it must be possible to record the tiny currents that flow through them, and in the early 1970s they set out to try. Their idea was to use a fine fluid-filled glass tube as a recording electrode. Because the tip of the tube was very small, when it was gently placed on the surface of the cell it was possible to isolate just one ion channel in the piece of membrane under the electrode tip. The tiny currents that flowed through the channel when it was open could then be detected. The technique is known as the 'patch-clamp' method because it records the current flowing though a minute patch of the cell membrane.

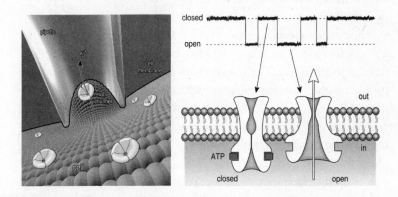

Left. The patch-clamp method showing how the glass electrode can isolate a single channel in a patch of the cell membrane, enabling the tiny currents that flow through the channel when it is open to be detected. *Right.* Single-channel current recording (above). When the channel flickers open (below), the brief current carried by the ions moving through it appears as a downward deflection of the trace. The channel shown below is closed when it binds the intracellular chemical ATP and opens when ATP unbinds.

It was some years before Neher and Sakmann succeeded, however. The difficulty was that they needed highly specialized equipment to amplify the very small signals and these were not only commercially unavailable, they had not even been invented. Thus they had to build the amplifiers themselves. Every time there was a new advance in technology they rebuilt their equipment and tried again. A crucial problem was that the tiny signal they were looking for was buried in the noise. Like the hiss you hear on an old radio, electric circuits (including biological ones) generate electrical noise. Neher and Sakmann tried many tricks to reduce the background noise and finally their perseverance paid off. Around 1974 they started to see single-channel currents emerge from the noise – these blips in the recordings appeared as tiny square pulses of current, generated as ions flowed through the pore each time the channel opened. At first they did not dare to report their results as they only saw currents under the most favourable conditions, but eventually, after a lot of work, they were sufficiently convinced to publish.

Their paper created quite a stir, but it was clear that this was not an easy technique to master, and few people tried to replicate their results immediately. The background noise also continued to be a problem, confounding measurements of small currents. For the next two years the scientists tried to improve the quality of the recordings and they became increasingly frustrated because nothing worked. And then, right out of the blue, they had a tantalizing glimpse of how good their recordings potentially might be. In previous experiments they had occasionally observed a dramatic reduction in noise – so much so that the current trace became a flat line – but, thinking that the tip of the electrode had become clogged up with debris, they immediately terminated the experiment (and inadvertently threw out the baby with the bathwater). However, on very rare occasions, a similar thing occurred when they were recording the usual little noisy blips and they were astonished to see the currents suddenly appear in startling clarity. What had happened, but they did not know it at the time, was that the cell membrane had sealed itself very tightly to the glass electrode, so eliminating almost all background noise. It afforded a step change in the resolution of the recording.

However, they could not reliably reproduce such perfect recordings until one Saturday afternoon in January 1980 it suddenly dawned on Neher that whenever he used a newly made electrode there was a higher chance of a seal forming. Elated, he called his colleague to tell him, 'I know how to get channels!' However, that was still not the end of the story: even with fresh pipettes, seals were not always forthcoming. Removing debris clinging to the cell membrane with enzymes, or using tissue cultured cells that naturally have very clean membranes, improved the success rate. But the final trick was to apply gentle suction to the electrode; that seemed to drag the cell membrane partly into the electrode and seal formation became more reliable. It had taken almost ten years to get there.

Real scientific breakthroughs occur far less often than one might imagine from reading the newspapers and they do not happen overnight – long years of hard grind are usually involved, as this story shows. But the perfected patch-clamp method was truly revolutionary. It quickly became clear that the technique was far more versatile than had been initially envisaged. The remarkable stability of the seal between the glass pipette and the cell membrane meant it was possible to rip small pieces of membrane off the cell without damaging them and so record channel activity in isolated membrane patches. The method could even be used to study all the many different types of cell in the body, which earlier technologies could not access because they injured the cell too much.

The paper from Neher and Sakmann's team detailing precisely how to obtain high-resolution recordings in all these different configurations electrified the scientific community and quickly became a classic. Almost overnight, everyone wanted to try patch-clamping. Neher and Sakmann generously threw open the doors of their labs and the whole world went to Göttingen to learn how to do it. Even then it was not easy, as you had to build the equipment yourself. I spent many weeks with a hot soldering iron in one hand, brushing away tears with the other hand, as I struggled with the complex electric circuits. Happily, this torture did not last long; within a few years it was possible to buy commercial amplifiers that worked perfectly (provided, that is, you had the money to do so).

Now that it was possible to see the channel's electrical signature all kinds of questions could be addressed. How many kinds of channel are there? What do they do? And how exactly do they work – what kind of molecular gymnastics do they perform when they open and shut, and how do they pick and choose which ions they let through?

A Genetic Toolkit

Almost at the same time as Neher and Sakmann were transforming our ability to see ion channels in action, a second scientific revolution was taking place. The blueprint to make every one of the proteins we possess is encoded in our DNA and the invention of new molecular biology techniques meant that it became possible to identify and manipulate the DNA sequence that codes for an individual protein. Proteins are formed from a linear string of amino acids but – like a bead necklace dropped on the floor – they fold up into far more complex shapes. Some of the protein may become embedded in the membrane while other bits sit inside or outside of the cell. The protein may even twist around so that part of its structure becomes inverted or, to paraphrase T. S. Eliot, its end may be at its beginning. The three-dimensional shape a protein adopts is critical: ion channels must provide a path for ions to flow through, signalling molecules have to dock snugly with their target receptors, structural proteins need to lock themselves tightly together. Sometimes several protein chains get together to produce an even more complex structure: potassium channels, for example, tend to be built of four similar subunits, which link up to form a central pore through which the ions move.

It is impossible (at present) to predict the three-dimensional structure of a protein simply from its amino acid sequence. Yet to fully understand how a channel operates it is important to have some idea of what it looks like. It was knowledge of the DNA sequence that provided the first step towards understanding the relationship between structure and function. Once the genetic code of a protein is known it is possible to change it, and to produce designer channels tailored to the question you wish to address. Wonder what a particular amino acid does? Simple: change it to a different one and see what happens. It sounds straightforward, and these days it is. Now that we know the complete sequence of the human genome (and that of many other species) it is possible to look up the DNA sequence of your favourite protein in an online database and order it up from a commercial company at a cost of around a thousand pounds: it arrives within a few days, an invisible drop on a tiny piece of filter paper. Back in the 1980s, however, things were not so easy. It was usually necessary to figure out the DNA sequence yourself and that could take time – many, many years in some cases.

The Needle's Eye

Nevertheless, the marriage of molecular biology and new electrical recording methods gradually began to address the problem of ion channel selectivity – of how channels choose between ions. It turns out that because similar charges repel one another and opposite charges attract, many channels use rings of charge at their entrances to exclude or enhance ion entry. For example, by using negative charges – which will encourage cations to enter, but repel anions – a channel can permit the passage of all positively charged cations, but exclude negatively charged anions. The crucial problem most ion channels must solve, however, is how to produce high selectivity without slowing down the rate at which ions move through the pore. And one of the most difficult questions to answer was how potassium channels allow potassium ions to permeate but not the much smaller sodium ions, which are also positively charged. It mystified scientists for years. The field had a skeletal idea, a cartoon model of how things worked, based on a plethora of functional experiments, but what was really needed was to couple this information with a structural understanding. Just what did a potassium channel actually look like? That puzzle was finally solved in 1998, when Rod MacKinnon achieved a stunning breakthrough: by growing crystals of a potassium channel protein and shooting X-rays at them, he was able to see each atom of the potassium channel for the first time. And potassium ions were caught 'in flagrante', trapped at various locations within the pore, so that the way they traverse the membrane could be seen in exquisite detail.

A slight figure with an elfin face, MacKinnon is one of the most talented scientists I know. He was determined to solve the problem of how channels worked and he appreciated much earlier than others that the only way to do so was to look at the channel structure directly, atom by atom. This was not a project for the faint-hearted, for nobody had ever done it before, no one really knew how to do it and most people did not even believe it could be done in the near future. The technical challenges were almost insurmountable and at that time he was not even a crystallographer. But MacKinnon is not only a brilliant scientist; he is also fearless, highly focused and extraordinarily hard-working (he is famed for working around the clock, snatching just a few hours' sleep between experiments). Undeterred by the difficulties, he simultaneously switched both his scientific field and his job, resigning his post at Harvard and moving to Rockefeller University because he felt the environment there was better. Some people in the field wondered if he was losing his mind. In retrospect, it was a wise decision. A mere two years later, MacKinnon received a standing ovation – an unprecedented event at a scientific meeting – when he revealed the first structure of a potassium channel. And ion channels went to Stockholm all over again.[2]

The X-ray structure showed in dazzling detail how the potassium channel works and how it is able to support a very rapid throughput of potassium ions, so fast it seems there is no barrier to ion movement at all, while at the same time excluding the smaller sodium ions. Potassium channels, it turns out, have evolved specialized 'selectivity filters', short regions where the pore narrows so much that permeating ions must interact with its walls. Simply put, this region is just wide enough for a potassium ion to squeeze through, but nothing larger can pass. The passage is so small, in fact, that potassium has to shed its coat of water molecules to squeeze through. In solution, all ions are clothed in bulky coats of water and it takes a lot of effort to remove them. Potassium is happy to shrug off its coat because the selectivity filter mimics the embrace of its watery jacket. Not so, sodium. Although sodium is small enough to slip through the pore when dehydrated, the effort required to strip off its water shell is too much – much greater than the energy supplied by the clinch of the selectivity filter – so it remains fully clothed. And with its coat on, sodium is simply too big to enter.

An Open and Shut Case

Ion channels are the gatekeepers of the cell. Arguably their most important property is that they open and close, so regulating ion movements, and, crucially, this gating is tightly controlled – by the binding of intracellular or extracellular chemicals, mechanical stress or changes in the voltage difference that exists across the cell membrane.

Nerve cells talk to one another by means of chemical messengers, known as transmitters, which interact with specialized ion channels in the membrane of the target cell. The transmitter binds to a specific site on the channel protein, fitting snugly into its receptor like a key in a lock. When it does so, it triggers a conformational change in the channel protein that opens the pore and enables ion flow. We still know little about how such shape shifting takes place, or how binding of a chemical at one site leads to a structural change in another part of the protein, which may be far distant. But this type of gating is very important, not just for transmitting information between cells but also because many medicinal drugs and many poisons influence channel activity (and thereby cellular functions) by binding to the same site as the native transmitter and either blocking, or mimicking, its action.

The South American Indian arrow toxin curare, for example, binds to ion channels involved in nerve-muscle transmission and prevents the action of the native transmitter, so producing paralysis. Conversely, the psychedelic drug LSD mimics the action of the transmitter serotonin, leading to overstimulation of certain neurones in the brain. My own particular favourite, the K_{ATP} channel, is closed by binding a breakdown product of glucose known as ATP; this is how glucose metabolism leads to the closing of the channel, and thus to insulin secretion. If the binding site is altered – for example, by a mutation like that which James carries – then ATP cannot bind, the K_{ATP} channel does not close, and insulin is not secreted. The result is diabetes.

'Voltage-dependent' gating requires that the channel is able to sense a change in the voltage field across the membrane. All cells have a potential difference across their membranes, the inside of the cell being about 70 millivolts more negative than the outside. When a nerve fires an electrical impulse this potential suddenly alters by about 100 millivolts, the inside of the cell briefly becoming positive with respect to the outside. A hundred millivolts may not sound a lot, but in fact it is, because the membrane is very thin. When the thickness of the membrane is taken into account the electric field experienced by the channel can be colossal: of the order of 100,000 volts per centimetre. Mains electricity in the UK is supplied at 240 volts and if you have ever been unfortunate enough to get a mains shock (I hope you never have, or will) you will appreciate the enormity of the electric shock that an ion channel experiences when a nerve fires an impulse. When put like this, it is less surprising that a voltage change can alter the protein conformation, switching its shape from one state to another. How the channel senses the voltage field has only been discovered in the last twenty-five years and the precise details are still the subject of heated debate.

In resting nerve and muscle cells, the voltage-gated sodium and potassium channels are held firmly shut by the negative membrane potential. They open only when the membrane potential becomes more positive and when this happens it triggers an electrical impulse. How this is achieved and the long and intricate work needed to unravel the story of how nerves and muscles work is considered in the next few chapters.

3

Acting on Impulse

I cannot see her tonight.
I have to give her up
So I will eat fugu.

Yosano Buson

During his voyage in the South Seas in 1774, Captain James Cook wrote the following account of the peculiar symptoms he experienced after sampling a strange and ugly fish: 'The operation of drawing and describing took up so much time, that it was too late, so that only the liver and roe were dressed, of which the two Mr. Forsters and myself did but taste. About three o'clock in the morning, we found ourselves seized with an extraordinary weakness and numbness all over our limbs: I had almost lost the sense of feeling, nor could I distinguish between light and heavy bodies, of such as I had strength to move; a quart-pot full of water and a feather being the same in my hand. We each of us took an emetic, and after that a sweat, which gave us much relief. In the morning, one of the pigs which had eaten the entrails was found dead. When the natives came on board, and saw the fish hang up, they immediately gave us to understand it was not wholesome food, and expressed the utmost abhorrence of it.'

It is possible that Cook and his crew had inadvertently eaten puffer fish. The liver, intestines, skin and ovaries of this fish contain a virulent poison known as tetrodotoxin, which acts by blocking the sodium channels in nerve and muscle cells. Consequently, nerve impulses and muscle contraction are inhibited. The victim typically dies of suffocation caused by paralysis of the respiratory muscles.

Cook was very fortunate that the amount he ate was insufficient to kill him.

Wiring the Body

Nerve fibres are used to transmit electrical signals around the body. What we generally refer to as a nerve is in fact a collection of many nerve fibres bundled together within a protective outer sheath, rather like a cable containing thousands of different telephone wires. Most nerves are located deep within our tissues to guard them against injury. The exceptions are the ends of the sensory nerves, which ramify throughout the outer layers of the skin, and the ulnar nerve that comes close to the surface of the skin at the elbow. This explains why a sharp blow to the elbow (the funny bone) produces a peculiar tingling pain that shoots down your arm: knocking the nerve excites it in the same way as a small electric shock.

Nerve cells are the building blocks of the nervous system, including the brain. They come in many shapes and sizes, but all consist of a cell body from which extend a number of fine, branched processes. Usually one of these processes is much longer than the others and is known as the nerve fibre or axon. It can be extremely long. The axons in your ulnar nerve, for example, run from your spinal cord to your fingers. The vagus nerve – the longest of the cranial nerves – runs from the brain to the stomach, and that of the giraffe can be more than 15 feet long. Yet despite its length, a single nerve fibre is very thin, with a diameter less than a tenth of that of a human hair.

Although nerve fibres are capable of conducting impulses in either direction, they usually only transmit them in one direction. Motor nerves conduct signals outwards from the brain and spinal cord to direct muscle contraction, whereas sensory nerves conduct information in the opposite direction, from our sense organs to the brain.

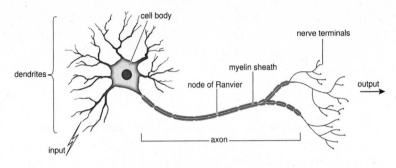

A typical neurone showing the axon and the numerous delicate branching dendrites that arise from the cell body, and the multiple finger-like processes at the axon terminal.

The cell body is the control centre of the nerve cell: it houses the nucleus where the genetic material (the DNA) is stored. Multiple short processes branch off the nerve cell body like the limbs of a tree, hence they are named dendrites, from the Greek *dendron*, meaning 'tree'. The dendrites receive numerous signals from other nerve cells and serve as first-line information processing centres, integrating all incoming information before passing it on to the cell body. Nerve cell bodies lie almost exclusively within the brain and spinal cord, where they are protected by a 'blood–brain barrier' which separates the blood from the cerebrospinal fluid bathing the brain and spinal cord. The brain serves as the command centre of the entire nervous system. It contains millions of nerve cells, each of which has multiple processes and multitudinous connections to other brain cells.

Acting on Impulse

Nerve cells transmit information by means of electrical signals known as nerve impulses or action potentials. These race along the nerve fibre at speeds of up to 400 kilometres per hour (250 miles per hour). The fastest nerves of all are those that are enveloped in an insulating myelin sheath. This is formed from layer upon layer of membranes of a specialized cell (the Schwann cell) that wraps itself tightly around the axon like the layers of a Swiss roll, or the paper layers enveloping a toilet roll tube. This insulating myelin sheath enables nerve fibres to conduct electrical impulses more rapidly. When it is damaged, nerve conduction is disrupted.

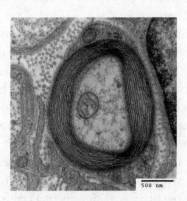

A myelinated nerve, showing the layers of insulating myelin wrapped around the nerve axon. The small organelle in the centre of the nerve is a mitochondrion, one of the cell's power plants.

Multiple Schwann cells are strung out along the length of the axon. Every few micrometres, adjacent Schwann cells are separated by small gaps known as the nodes of Ranvier, which allow the naked nerve membrane to contact the extracellular fluid. Because the myelin sheath is such a good insulator, it is only at the nodes that electric current can flow from the nerve cell to the extracellular fluid. The nodes thus serve as repeater stations, boosting the action potential and enhancing its speed. In effect, the nerve impulse travels faster in myelinated nerves because its leading edge leaps forward one node at a time. This explains why myelinated nerves conduct action potentials much faster than unmyelinated nerve fibres.

A dramatic example of the crucial importance of myelin is afforded by Guillain-Barré syndrome. This rare autoimmune disease usually begins with tingling and weakness in the feet, and is followed with frightening speed by paralysis of the lower limbs, then the hands and arms, and subsequently the chest muscles, so that the victim is unable to breathe and must be kept alive by artificial respiration. Ultimately, almost all the nerves may be affected, including those of the face, so that the person may be unable to speak and can only communicate by eye blinks. In the worst case you can go from normal nerve function to near-total paralysis within a day.

Guillain-Barré syndrome is caused by antibodies produced by the body against foreign proteins that for unknown reasons also attack its own tissues, in a form of cellular friendly fire. This leads to loss of myelin and destruction of the nerve sheath, which prevents impulse conduction. The brain and spinal cord are spared because the antibodies cannot cross the protective blood–brain barrier that surrounds them, and are thus barred from reaching the myelinated fibres within the brain. Fortunately the paralysis is usually not permanent and once the antibodies have been cleared from the system, the myelin grows back. But it is a slow process, taking about a centimetre a day, and in a tall person it can be well over a year before some muscles are fully reinnervated. In many cases, full function is never regained.

Similarly, multiple sclerosis is caused by a gradual and inexorable autoimmune attack on the myelin sheath, which results in progressive impairment of nerve conduction and eventually loss of coordination and difficulty in walking. It can also cause blindness, due to damage to the optic nerves. One of its most celebrated victims was the gifted and charismatic young British cellist, Jacqueline du Pré. When she was only twenty-six, she started to lose sensitivity in her fingertips and soon she was unable to feel the strings of her cello at all. She ceased performing two years later.

Listening to Nerves Talk

We humans have been digital for years. Long before computers were even dreamt of, impulses were being sent in digital code along our nerve fibres. Action potentials are said to be 'all-or-none', as their amplitude is constant and independent of the strength of the stimulus that evokes it: instead, increasing the stimulus intensity provokes a higher frequency of action potentials. A good analogy is with a machine gun. Press the trigger hard enough and the gun fires, but if the stimulus fails to reach a certain threshold level no bullets (or action potentials) are fired. Furthermore, much like a stream of bullets fired from a machine gun, information is transmitted along nerves as a volley of identical action potentials, with stronger stimuli eliciting a greater number of spikes. This frequency coding has significant advantages. It ensures, for example, that electrical impulses are transmitted over long distances without the information becoming garbled or the signal strength decaying.

In order to study how nerve impulses are generated and propagated, detectors sensitive enough to pick up the tiny brief electrical signals are needed. What tantalized early investigators like Galvani was that although they could easily detect the result of the nerve impulse – the twitch of a frog's muscle – they were quite unable to record it electrically. By the middle of the nineteenth century specialized instruments known (in a nod to Galvani) as galvanometers had been developed. Using such equipment, many investigators deduced that nerves and muscles do indeed generate their own electrical signals, but they were still unable to measure them accurately. Frustratingly, if their instruments were sensitive enough they were too slow, and if they were fast enough they were not sufficiently sensitive. Neuroscientists had to await the invention of the thermionic (triode) valve, originally developed for radio communication, in order to build amplifiers capable of detecting nerve impulses accurately.

Edgar Adrian and Keith Lucas carried out pioneering experiments with this new instrumentation, using it to amplify the tiny electrical signal produced by a nerve fibre around 2,000 times. Adrian was a strong advocate of the importance of technology, believing that 'the history of electrophysiology has been decided by the history of electric recording instruments'. He practised what he preached and his laboratory 'contained the most glorious clutter even seen'. He also said that his results 'didn't involve any particular hard work, or any particular intelligence on my part. It was one of those things which sometimes just happens in a laboratory if you stick apparatus together and see what results you get.' Adrian's protégé Alan Hodgkin later dryly remarked that when most people 'stick apparatus together and look around they do not make discoveries of the same importance as those of Adrian'. Interestingly, Hodgkin was one of the few who did.

Adrian began as an assistant to Keith Lucas while still a student at Cambridge in 1912. Lucas's laboratory was memorable for being sited in a tiny dark dank cellar that flooded every time it rained, so that the scientists were forced to walk about on duckboards in wet weather (not the best environment for electrical experiments and one that would surely be banned on health and safety grounds today). Lucas set Adrian an exciting challenge – to investigate the conduction of the nerve impulse. Earlier experiments had already provided some support for the idea that nerve fibres fire fully or not at all, but there remained room for doubt; however, before they were able to settle the case one way or another, their research was interrupted by World War I. Indeed, Lucas's career came to a permanent halt as he was killed in a mid-air collision while carrying out instrument tests for the Royal Air Force. After the war was over, Adrian took over his mentor's laboratory in Cambridge. By painstakingly dissecting a single nerve fibre away from its neighbours in a nerve bundle, he discovered that when he stimulated the nerve, it generated a series of tiny electrical impulses of regular amplitude but variable frequency; the greater the stimulus strength, the higher the discharge rate. In other words, the intensity of a sensation is proportional to the frequency of sensory nerve impulses.

Adrian remarked that constant reminders of this fact come unexpectedly, citing a memorable experiment in which he had placed some electrodes on the optic nerve of a toad. 'The room was nearly dark and I was puzzled to hear repeated noises in the loudspeaker attached to the amplifier, noises indicating that a great deal of impulse signalling was going on. It was not until I compared the noises with my own movements about the room that I realized that I was in the field of vision of the toad's eye and that it was signalling what I was doing.'

Chance and Good Fortune[1]

By the middle of the last century it was appreciated that nerves and muscles transmit information in the form of electrical impulses, but exactly how the nerve impulse was generated and propagated along the nerve fibre was still a mystery.

The pioneering experiments that led to the solution of this problem were carried out using single nerve fibres of the squid, giving this animal a special place in the hearts of physiologists. It was John Zachary Young, a scientific polymath affectionately known as JZ, who discovered that the common squid (*Loligo forbesii*) has a nerve fibre large enough to be seen with the naked eye. Tall and distinguished, with a thick shock of silver hair and an infectious enthusiasm, JZ was a memorable individual. Every summer he would disappear to Plymouth or Naples to pursue his studies of octopus and squid, and it was while he was there that he first observed that the squid mantle is innervated by enormously thick nerve fibres. These giant cells conduct nerve impulses very fast and are responsible for initiating the rapid jet-propelled escape of a squid when confronted by an enemy. They also provided scientists with an invaluable preparation for studying how nerve impulses are generated, and an excellent excuse for spending large amounts of time at the seaside. Two marine laboratories where fresh squid were obtainable became particularly popular: the Marine Biological Laboratories at Plymouth in England and those at Woods Hole on Cape Cod in the USA.

The huge size of the giant squid axon – between a half and one millimetre in diameter – meant that it was possible to insert an electrode inside the axon and for the first time measure the voltage difference between the inside and the outside of the cell. This was achieved by a famous partnership between two young Cambridge scientists, Alan Hodgkin and Andrew Huxley, in early August 1939.[2] For Huxley, still a medical student, this was his first taste of research. They carefully dissected out a single giant nerve fibre, hung it vertically from a hook, and threaded a thin silver wire (protected by a glass capillary) down its length, steering it straight down the middle of the axon without touching the sides with the help of a small mirror. A second electrode was placed in the seawater surrounding the axon. This enabled them to measure the voltage difference between the inside and outside of the cell simply by measuring the voltage difference between the two electrodes.

When they did this, they found that the inside of the resting nerve cell was around 50 millivolts more negative than the outside. This was not entirely unexpected, as a negative resting potential had already been predicted. It is produced by a tiny leak of positively charged potassium ions out of the cell at rest, as described in the previous chapter. The big surprise was when they stimulated the nerve with a

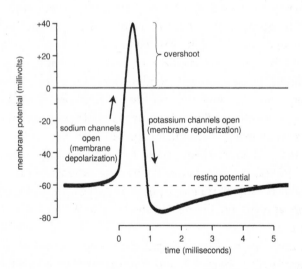

Action potential: showing the negative resting membrane potential and the transient positive overshoot that occurs when the nerve cell fires an impulse.

small electric shock and evoked a nerve impulse, for they found this produced a transient reversal in the voltage difference across the membrane such that the inside of the cell became almost 50 millivolts more positive than the outside. This 'overshoot' in potential was remarkable for being completely contrary to existing dogma and it necessitated a rethink in how nerves might work.

Hodgkin and Huxley recorded the first action potentials on 5 August 1939 and were tremendously excited about their discovery. They quickly dashed off a brief note to the journal *Nature*, but with no explanation of their findings. Three weeks later, on 1 September, Hitler marched into Poland, Britain declared war on Germany and the scientists had to abandon their experiments for eight years. It must have been immensely frustrating, but they had little time to brood on it, as they were soon engaged in the more pressing problem of how to win the war.

During the first few months of hostilities, Hodgkin tried to write up their results as a full paper, but he did not get very far as his war duties kept him very busy; and by 1940 'the war had gone so disastrously, and the need for centimetric radar was so pressing' that he 'lost all interest in neurophysiology' and worked flat out on radar. This necessitated a considerable investment of time in learning the necessary physics, but it was not without its excitements, for Hodgkin was soon engaged in developing a short-wave airborne radar system for night-fighter aircraft that was capable of detecting enemy bombers in the dark. This involved much in-flight testing of the prototypes in unpressurized aircraft, and the early high-voltage equipment was prone to arc in the rarefied air at altitude, setting fire to the instruments and filling the plane with smoke. Huxley was also kept very busy working on the application of radar to anti-aircraft naval gunnery.

Although the British scientists were unable to continue their experiments during the war, work did not stop in the United States. Kenneth Cole (known as Kacy) and his colleague Howard Curtis also started to record action potentials from squid axons at Woods Hole. Unfortunately, some of their results were misleading.

They illustrated their paper with a 'typical' recording, but like many scientists they did not show a truly representative trace, but rather the one that they considered their best – and they picked the biggest. But bigger is not always better and unfortunately this recording was flawed. In retrospect, Cole suggested their equipment might have been poorly adjusted because their action potential overshot the zero potential by almost 100 millivolts. This giant action potential did not fit with any known theory and it held back understanding of how the nerve impulse works for some time: a cautionary tale that serves to remind scientists that a 'typical record' really should be typical.

Taming the Axon

After the war was over, Hodgkin and Huxley teamed up again and in 1945 finally wrote up the results of their 1939 experiments in detail. They produced a full-length paper with four possible explanations of their results – all of which were wrong, as they later acknowledged. What was needed was more experiments. However, it was not easy to get the Plymouth laboratories going again as they had been partly demolished in air raids, squid were in short supply and as Hodgkin remarked, he 'had forgotten much of the technique'. When they were finally able to restart experiments, in 1947, Huxley was on his honeymoon, so Hodgkin enlisted the help of Bernard Katz, a young refugee from Nazi Germany.

During the war years, Hodgkin and Huxley had become convinced that the action potential must be caused by a transient increase in the permeability of the nerve membrane to sodium ions, and Hodgkin was eager to test this idea. To his satisfaction, he found he could record impulses when the axon was bathed in normal seawater but not when the sodium ions in the seawater were replaced with another ion. Thus it seemed that a current carried by sodium ions moving from the external solution into the axon might underlie the overshooting nature of the action potential. Parenthetically, this current is caused by the opening of sodium channels in the axon membrane but at the time no one, including Hodgkin and Huxley, knew that ion channels even existed.

The big problem in understanding precisely how nerves work was that the action potential was all-or-none. Nothing happened until the electrical stimulus exceeded a certain threshold and then everything happened all at once, with the membrane potential suddenly and explosively switching from its resting level to one some 100 millivolts more positive and then rapidly switching back again. What was needed was some way to prevent the stimulus-induced change in membrane potential and make the voltage stand still, so that the currents associated with a potential change imposed by the experimenter could be measured in isolation. This was achieved with the help of an ingenious device known as a voltage clamp. The way it worked was to inject a current that was equal in amplitude but opposite in direction to that which flowed across the membrane. That way the transmembrane potential did not alter, as the membrane current was cancelled out. Furthermore, the magnitude of the current flowing across the membrane was directly proportional to that injected by the voltage clamp, enabling precise measurements of the currents underlying the action potential. It was a brilliant solution to the problem.

The voltage clamp was invented independently by Hodgkin and Katz in Plymouth, and by Cole and George Marmont in Woods Hole. The Americans were more advanced in its development and were the first to conduct experiments with the voltage clamp (a term that Cole disliked), in 1947. Cole informed Hodgkin of their experiments and when Hodgkin visited Cole at Woods Hole in March 1948 they swapped experimental details. Hodgkin quickly realized that Cole's apparatus was rather better than his own. On his return to England, Hodgkin and Huxley modified their equipment, incorporating improvements on Cole's method, and in one short month in August 1949 they obtained all the results that were needed to show how nerves work. The secret to their success lay in the sophisticated design of their experiments and an approach quite distinct from that of Cole.

Cole was astonished at the speed of their progress, commenting, 'Hodgkin and Huxley went ahead with amazing speed [. . .] I had occasional reports from them. But again I did not appreciate the beautiful simplicity of the fundamental concepts and the spectacular detail and successes of [their analyses. : . .] It was only after Hodgkin had sent me drafts of their manuscripts [. . .] that I began to understand what had grown from my simple idea to tame the squid axon.' The phrasing of the latter sentence, together with his statement that through 'free exchange of methods and results, they [i.e. his rivals] were able within a year to repeat all my work with very considerable improvements', contains hints of the inner turmoil that the Cambridge scientists' success must have generated.

Hodgkin and Huxley's elegant experiments revealed precisely how the nerve generates an electrical impulse. The action potential is caused by an initial increase in the permeability of the membrane to sodium ions. This is produced by the opening of sodium channels, which allows positively charged sodium ions to rush into the nerve cell and drive the membrane potential positive (depolarization). Less than a millisecond later, the potassium channels open, permitting potassium ions to exit the nerve and return the membrane potential to its resting level (repolarization). Together, these opposing ion fluxes generate a transient change in voltage that constitutes the nerve impulse.

Calculated Progress

Having measured the amplitude and time course of the sodium and potassium currents, Hodgkin and Huxley needed to show that they were sufficient to generate the nerve impulse. They decided to do so by theoretically calculating the expected time course of the action potential, surmising that if it were possible to mathematically simulate the nerve impulse it was a fair bet that only the currents they had recorded were involved. Huxley had to solve the complex mathematical equations involved using a hand-cranked calculator because the Cambridge University computer was 'off the air' for six months. Strange as it now seems, the university had only one computer at that time (indeed it was the first electronic one Cambridge had). It took Huxley about three weeks to compute an action potential: times have moved on – it takes my current computer just a few seconds to run the same simulation. What is perhaps equally remarkable is that we often still use the equations Hodgkin and Huxley formulated to describe the nerve impulse.

Three years after finishing their experiments, in 1952, Hodgkin and Huxley published their studies in a landmark series of five papers that transformed forever our ideas about how nerves work. The long time between completing their experiments and publication seems extraordinary to present-day scientists, who would be terrified of being scooped by their rivals. Not so in the 1950s – Huxley told me, 'It never even entered my head.' In 1963, Hodgkin and Huxley were awarded the Nobel Prize. Deservedly so, for they got such beautiful results and analysed them so precisely that they revolutionized the field and provided the foundations for modern neuroscience.

The Scramble for Squid

Hodgkin and Huxley's experiments generated much excitement and led to an annual migration of scientists to the marine laboratories at Plymouth and Woods Hole. As squid are migratory and the scientists had teaching duties this inevitably turned into 'summer camp for scientists', and – particularly at Woods Hole – led to a hothouse of experiments and ideas. Squid were in short supply and the best squid were keenly fought for so that a pecking order quickly developed. By the mid-1960s the hectic scramble for squid was so considerable it motivated some scientists to find a place to work in the winter, and Montemar, near Valparaiso in Chile, provided the perfect place. There was an added bonus, as the Chilean squid – and their axons – are much larger.

Although many other cell types are used to investigate the mechanism of the nerve impulse today, including mammalian brain cells, the squid axon remains a valuable preparation for scientific study. In Plymouth during the 1940s the few squid caught were so mangled by the trawlers' nets that they did not survive long once they arrived back at the lab, which meant that experiments had to be conducted immediately. Because the boats did not return until late afternoon, this usually meant working throughout the night. Consequently, Hodgkin and Huxley spent their mornings catching up on sleep and planning experiments. This was also true when I visited Woods Hole in the 1980s, when many scientists crawled into bed around 4 a.m. after a hard night at the bench. In Chile today, many squid are caught by rod and line and consequently suffer less damage. But their huge size means they are less easy to keep in holding tanks, so scientists still face the night shift.

I vividly recall from the time I spent at Woods Hole that the axons which generated the best results were commemorated in a most singular fashion. At the end of the experiment, they were flicked onto the ceiling of the laboratory, which eventually acquired a pattern of dried-out squiggles, somewhat reminiscent of a Jackson Pollock painting. Only the very best axons, however, were 'sent to Heaven'.

Fire!

Sodium and potassium channels that open in response to changes in the voltage gradient across the cell membrane are the keystone of electrical signalling in our brain, heart and muscle. In resting nerve cells, both kinds of channel are tightly shut. When the nerve is stimulated, first the sodium channels and then, with a short delay, the potassium channels swing into action producing a transient change in membrane potential – the nerve impulse. But what triggers the whole thing off?

Crucially, the sodium and potassium channels that are involved in the action potential are sensitive to voltage and they open if the membrane potential is made more positive (depolarized). This is exactly what happens when a nerve cell is excited by an incoming signal from another nerve cell, or by an externally applied electric shock. The larger the change in membrane potential this produces, the more sodium channels open and the more sodium ions flood into the cell. You may recall that Ohm's law dictates that a change in current will produce a concomitant change in voltage. In a nerve cell the sodium current drives the voltage more positive, which opens more sodium channels, which makes the membrane more positive, which opens more channels, and so on and so on in a positive feedback cycle. This explains the explosive, all-or-none nature of the action potential.

Two things return the membrane potential to its resting level. The sodium channels do not remain open forever at positive membrane potentials, but eventually close, a process known as inactivation. Secondly, the potassium channels open, so that potassium ions rush out of the cell, restoring the charge imbalance and sending the potential negative once more. It's just as well that the potassium channels open later than the sodium channels because if they opened at the same time the sodium and potassium currents would cancel each other out and there would be no nerve impulse, and no thoughts or actions.

Terrible Stuff

The importance of sodium and potassium channels in generating the nerve impulse is demonstrated by the fact that a vast array of poisons from spiders, shellfish, sea anemones, frogs, snakes, scorpions and many other exotic creatures interact with these channels and thereby modify the function of nerve and muscle. Many are highly specific and target a single kind of ion channel. Which brings us back to Captain Cook and the puffer fish.

The tetrodotoxin contained in the liver and other tissues of this fish is a potent blocker of the sodium channels found in your nerves and skeletal muscles. It causes numbness and tingling of the lips and mouth within as little as thirty minutes after ingestion. This sensation of 'pins and needles' spreads rapidly to the face and neck, moves on to the fingers and toes, and is then followed by gradual paralysis of the skeletal muscles, resulting in loss of balance, incoherent speech, and an inability to move one's limbs. Ultimately, the respiratory muscles are paralysed, which can be fatal. The heart is not affected, as it has a different type of sodium channel that is far less sensitive to tetrodotoxin. The toxin is also unable to cross the blood–brain barrier so that, rather horrifyingly, although unable to move and near death, the patient remains conscious. There is no antidote and death usually occurs within two to twenty-four hours. In 1845, the surgeon on board the Dutch brig *Postilion*, sailing off the Cape of Good Hope, observed that two seamen 'died scarcely seventeen minutes after partaking of the liver of the fish'. However, victims can recover completely if they are given artificial respiratory support until the toxin has washed out of the body – which takes a few days.

In Japan, the puffer fish is known as fugu, and is considered a

Hiroshige's 'Amberjack and Fugu'. The puffer fish (fugu) is the smaller fish.

great delicacy. Unfortunately, the fish is expensive in more ways than one, as unless it is carefully prepared the flesh can be toxic, and every year several people die from tetrodotoxin poisoning. Most incidents arise from fishermen eating their own catch. Restaurant casualties are far rarer because all fugu chefs must now be specially trained and licensed, which involves passing a rigorous test. Nevertheless, it occasionally happens. One of fugu's most celebrated victims was the famous Japanese kabuki actor Bando Mitsugoro who died after eating it in 1975: he had demanded four servings of the liver, which is especially dangerous, and the restaurant had felt unable to refuse such a distinguished customer. Perhaps this is why fugu is forbidden to the Emperor of Japan. Properly prepared, the fish is supposed to cause a very mild intoxication and produce a stimulating, tingling sensation in the mouth. On the single occasion when I tried it myself, I found it rather insipid: it was the spice of danger that enlivened the dish.

Not all cases of fugu poisoning are due to the deliberate ingestion of the fish. In 1977, three people died in Italy after eating imported puffer fish mislabelled as anglerfish. Ten years later, two people in Illinois developed symptoms resembling those of tetrodotoxin poisoning after eating soup made from imported frozen 'monkfish'. Analysis by the FDA confirmed the presence of the toxin and triggered a mass recall of all sixty-four crates of the imported product. Claims lawyers instantly leapt into action. Poisoning from commercially cooked shellfish is also worryingly common in China and Taiwan: between 1997 and 2001 three hundred people were intoxicated and sixteen died.

A wide variety of animals contain tetrodotoxin, from reef fish, crabs and starfish to marine flatworms, salamanders, frogs and toads. Most use it as a biological defence, but some, like the deadly blue-ringed octopus, package it in venom to poison their prey. It was a mystery why so many different kinds of animal should make tetrodotoxin until it was discovered that it is actually made by a bacterium (*Psuedoalteromonas tetraodonia*) that the animal eats or harbours within its intestine. Puffer fish reared in the absence of such bacteria do not contain tetrodotoxin. Whether such fish, in which the element of Russian roulette is removed, will be as highly prized by aficionados as the native fish is an interesting question.

The fictional British agent James Bond (007) appears to have a special attraction for tetrodotoxin, for he has been poisoned with it on no less than two occasions. *From Russia with Love* ends on a moment of high drama when the SMERF agent Rosa Klebb kicks him with a poison-filled spike mounted on the tip of her boot and he is left to die. Bond, of course, is invincible and the next novel (*Dr No*) begins with him recovering from what we learn is a near fatal dose of tetrodotoxin – 'terrible stuff and very quick'. He survives only because his companion administered artificial respiration until medical help arrived. Bond also has an encounter with a blue-ringed octopus, whose bite is laced with tetrodotoxin, in the film *Octopussy*. As ever, 007 emerges from these incidents shaken but not stirred.

Red Tides and Suicide Potions

When conditions are right, spectacular blooms of the algae *Alexandrium* can occur that turn the sea the colour of blood. The deleterious effects of such red tides have been known for centuries and the biblical account of one of the great plagues of Egypt paints a vivid picture: 'all the waters that were in the river were turned to blood. And the fish that was in the river died; and the river stank and the Egyptians could not drink of the water of the river.' Such red tides are composed of millions of minute algae, known as dinoflagellates, which produce a number of virulent neurotoxins, including saxitoxin. Like tetrodotoxin, saxitoxin blocks sodium channels. Filter-feeding molluscs like mussels and clams may ingest the dinoflagellates, thereby concentrating the toxins they produce, and creatures that feed on them may, in turn, be poisoned. As much as 20,000 micrograms of saxitoxin per 100 grams of tissue (250 times the legally allowed limit) has been recorded in Alaskan mussels: at this level, consumption of a single mussel can kill you. Even more frighteningly, a single green shawl crab from the Great Barrier Reef can contain enough toxin to kill 3,000 people. Dinoflagellates are most abundant in spring and summer, due to the higher sunlight levels and warmer waters, which may be the origin of the old adage 'Do not eat shellfish unless there is an R in the month'.

In developed countries shellfish poisoning is very rare, due to intensive surveillance programmes and stringent regulations which ensure that, once detected, the affected areas are quarantined and shellfish sales prohibited. In the last decade, seasonal outbreaks of paralytic shellfish poisoning (mainly due to saxitoxins) have led to temporary bans on the sale of shellfish from waters around the world. The Alaskan shellfish industry has been radically affected as the butter clam is toxic for large parts of the year. But while commercial seafood is safe, this is not necessarily the case for shellfish that people collect themselves. Between 1973 and 1992 there were 117 cases of paralytic shellfish poisoning in Alaska, 75 per cent of them between May and July. Fortunately only one person died, but many required hospitalization. The most dramatic outbreak in recent years happened in 1987 in Guatemala, when 187 people were affected by eating clams and 26 of them died.

Tetrodotoxin and saxitoxin are molecular mimics. Each physically plugs the external mouth of the sodium channel pore, is almost equally potent at inhibiting channel function, and produces similar physiological responses. Both are also valuable research tools because they block sodium channels rather specifically, leaving most other channels untouched. Tetrodotoxin is routinely used in scientific studies today to block sodium channels and enable other channels to be studied in isolation. Its action was discovered by Toshio Narahashi in 1962, working round the clock throughout Christmas and New Year with John Moore and William Scott. Narahashi recalls that the reviewer of their manuscript jotted down a request for some of the toxin at the bottom of his report. It was to be the first of many such requests.

By now you might be wondering why butter clams are not affected by the saxitoxin they contain and why puffer fish swim happily around, despite high tetrodotoxin levels. The answer is that the affinity of their own sodium channels for the toxin is dramatically reduced because evolution has changed one or more of the amino acids in the toxin-binding site. A similar mutation is found in the cardiac sodium channel, which helps explain why your heart continues to beat even when your respiratory muscles are totally paralysed.

Saxitoxin was exploited by US agents engaged in covert government operations both as a suicide and an assassination agent. It has the advantage that it is highly poisonous so that only tiny amounts (which can be easily concealed) are needed, and it is faster and more effective than cyanide. Because it is stable, water soluble and about a thousand times more toxic than synthetic nerve gases such as sarin, saxitoxin (known as agent SS or TZ) was also stockpiled by the US government as a chemical weapon. It was extracted from thousands of butter clams laboriously collected by hand in Alaska. In 1969/70, President Nixon halted the US biological weapons programme and ordered existing stocks to be destroyed, in accordance with a United Nations agreement. But five years later, Senator Frank Church, Chair of a Select Committee on Intelligence investigating the CIA, discovered that a middle-level official had failed to do so. About 10 grams of the toxin, enough to kill several thousand people, still remained in downtown Washington in direct violation of the presidential order. It had been packed into two one-gallon cans and stored in a small freezer under a workbench, which must have caused some consternation to its discoverer.

This information interested Murdoch Ritchie, of Yale University School of Medicine, as he realized that the toxin could be of considerable value to scientists studying how nerves work. He immediately wrote to Church requesting that it should not be incinerated. To his surprise, the CIA offered Ritchie the entire supply, with the proviso that he organize its distribution to the scientific community. Ritchie quickly realized that safeguarding the stockpile would be an enormous responsibility. Moreover, as the supply was limited and the demand was likely to be considerable, he might 'be forced to ration it, or even deny some applications, and would surely make enemies'.

Wisely, Ritchie recommended it be given to the National Institutes of Health for distribution. The outcome was a happy bonus for ion channel research.

Saxitoxin has always been hard to obtain because of its colourful history, and the first laboratory synthesis of saxitoxin (in 1977) led to even more stringent controls. It is now listed in Schedule 1 of the Chemical Weapons Convention. In contrast, for many years tetrodotoxin could be routinely ordered from suppliers of laboratory chemicals. Since 11 September 2001, however, things have tightened up worldwide. Researchers can only hold tiny amounts of the toxin, and all stocks must be registered. They are also carefully monitored, as I discovered recently myself when we received an unsolicited visit from the anti-terrorist branch of the British police to check on our tetrodotoxin supplies.

The Queen of Poisons

Not all sodium channel toxins work by blocking flux through the pore. Some produce equally devastating effects by locking the channels open, resulting in overstimulation of nerve and muscle fibres. One of the most potent is aconite, which has been used as a murder weapon for centuries. A recent victim was Lakhvinder Cheema, who came home, took some leftover vegetarian curry out of the fridge and heated it up for himself and his fiancée, Gurjeet. They sat down to eat their dinner, chatting happily about their forthcoming wedding. But not for long. Ten minutes or so later, Lakhvinder found his face becoming numb and very quickly both he and Gurjeet started to go blind, became dizzy, and lost control of their arms and legs. They called for an ambulance, but Lakhvinder died within the hour and Gurjeet was left fighting for her life. She survived only because she had eaten less curry. The dish had been laced with aconite by Lakhvinder's jealous ex-mistress Lakhvir Singh, who had slipped into his flat when he was out.

Aconite, or more correctly aconitine, is colloquially called the

Queen of Poisons and it comes from monkshood (wolfsbane), a pretty plant with a tall spike of blue helmet-shaped flowers often grown in gardens. In Greek mythology it is said to have sprung from the saliva that dripped from the ravening jaws of Cerberus, the three-headed dog that guards the gates of Hell. Its poisonous properties have intrigued writers for centuries and it features in numerous literary works, including Oscar Wilde's story 'Lord Arthur Savile's Crime' and James Joyce's novel *Ulysses,* in which Rudolph Bloom dies as a 'consequence of an overdose of monkshood (aconite) self-administered in the form of a neuralgic liniment'. Similarly, Ovid relates how Medea tried to kill Theseus by lacing his wine with aconite. People have also died from accidental ingestion of aconite, one of the best-known victims being the Canadian actor Andre Noble. Because the toxin is absorbed through the skin, even picking the plant without wearing gloves may cause symptoms. As Keats advises, one should not 'twist Wolfs-bane, tight-rooted, for its poisonous wine'.

Another potent sodium channel opener is batrachotoxin, which is secreted from glands on the backs of the vividly marked yellow and black 'poison dart' frogs of South and Central America. It is collected by the Chocó Amerindians, who use it to tip their blow darts. Like aconitine, there is no known antidote. Batrachotoxin is not made by the frog itself but instead acquired from the beetles on which they feast – although whether the beetles make the poison themselves, or, in turn, get it from something they eat is still uncertain. Poison dart frogs are not the only creatures to steal beetle batrachotoxin to use as a defence. The New Guinea hooded pitohui, resplendent in showy red and black plumage, does so too. Its feathers and skin are laced with batrachotoxin as the biologist John Dumbacher discovered to his cost when he sucked his fingers after being scratched by a bird he was trying to free from a net: both his fingers and lips quickly developed a tingling sensation and then went temporarily numb. His local guides informed him that pitohui were 'rubbish birds' and well known to be poisonous.

Equally fascinating is grayanotoxin, which also locks sodium channels open. It is produced by some species of rhododendron and becomes concentrated in the honey of bees that feed on the flowers' nectar. Consumption of affected honey causes 'mad honey syndrome' and has been poisoning us for centuries. One of the earliest accounts is by Xenophon, who relates that during an expedition near Trabzon (on the Black Sea) in 401 BC, 'The number of bee hives was extraordinary, and all the soldiers that ate of the honey combs, lost their senses, vomited, and were affected with purging, and none of them were able to stand upright; such as had eaten only a little were like men greatly intoxicated, and such as had eaten much were like mad-men, and some like persons at the point of death. They lay upon the ground, in consequence, in great numbers, as if there had been a defeat; and there was general dejection.' Something of an understatement, one imagines. But no one died and by the next day all had recovered their senses. The symptoms suggested to military commanders in classical times that mad honey might prove an effective biological weapon, if it were strewn in the path of the opposition, and in 67 BC three Roman cohorts (about 1,440 soldiers) under the command of General Pompey were slaughtered by enemy forces while incapacitated from consuming mad honey. Today, mad honey poisoning rarely occurs, for only certain varieties of plants make the toxin and commercial honey is blended so that any toxin present is diluted out. A few cases occasionally occur in the Black Sea region of Turkey where, rather improbably, mad honey is thought to enhance sexual performance. Fortunately, it is usually not fatal.

Sodium channel toxins can make valuable insecticides if they can be targeted to ion channels found specifically in insect nerves. One of the most famous is dichlorodiphenyltrichloroethane, otherwise known as DDT. It opens sodium channels in insect nerves, but not those in mammalian nerves, which are genetically and structurally distinct. Activation of the sodium channels causes insect axons to fire impulses spontaneously, which leads to muscle spasms and eventually to death. DDT played an important part in controlling the spread of typhus and malaria during and after World War II, but its efficacy gradually diminished as insects became resistant to its effects. This was because strong evolutionary pressure led to genetic mutations that altered the binding site for DDT on the sodium channel (thus preventing its action) becoming widespread. Use of DDT also became increasingly controversial after the publication of Rachel Carson's book *Silent Spring*, which linked DDT and other pesticides to a marked decline in songbird numbers in the United States. Although DDT does not open avian or mammalian sodium channels, it has other effects: in birds, for example, it causes thinning of the eggshell, leading to breakage and fewer hatchlings.

Sodium Rules

It is evident from this string of cautionary tales and grisly stories that Nature has evolved a vast number of toxins that target sodium channels – far more than interact with other types of ion channel. One reason for this may be that sodium channels are specialized for fast conduction of nerve and muscle impulses. Block them, and your prey will be swiftly paralysed and more easily caught.

The many toxins that target ion channels have also been of considerable value to scientists struggling to understand how nerves work. As they are often highly specific in their targets they can be used to dissect out the contribution of an individual channel to the electrical activity of a given cell. Today, toxins can be simply bought from a specialist company. In the past, it was a different matter and the scientists not only had to purify the toxins themselves, but often had to collect the animals that produced the venom as well.

The toxin hunters were an adventurous bunch. Some spent a month each year gathering cone snails in the Red Sea, alternately diving for snails and extracting the venom; it might sound like a holiday, but it was actually hard work, to say nothing of the added frisson provided by the need for eternal vigilance to avoid being stung. Other intrepid investigators travelled to North Africa, where they would venture out into the desert at night. Under a vast sky that glittered with stars, the ground appeared a deep inky black, free from danger. Switch on an ultraviolet lamp, however, and the many scorpions swarming around the scientists' feet leapt into view, for scorpions fluoresce brightly in ultraviolet light. Thousands were carefully packed into large milk churns and taken back to France, and a whole week each month was devoted to milking them for their venom. This operation requires extreme care, but an experienced researcher can avoid being stung by grabbing the scorpion close to the tip of its tail. The scorpion reacts by secreting a drop of venom from its sting, which is then carefully collected. It is a very time-consuming process because venom from as many as 150,000 scorpions is needed to isolate enough toxin for experiments.

The cornucopia of sodium channel poisons illustrates just how crucial sodium channels are for nerve and muscle function. Their special properties of voltage sensitivity and selective permeability to sodium ions are essential for the generation and conduction of nerve impulses, and the transmission of information along nerve axons, such as those that travel along motor axons from your brain and spinal cord to your muscles. What happens when those impulses reach the nerve terminals and how they then excite the muscle fibre is considered next.

4

Mind the Gap

the leap
thought makes at the synaptic gap.

Brian Turner, 'Here, Bullet'

Botox is the latest tool in the armoury of the cosmetic surgeon and is used by film stars and ordinary people alike to smooth out the wrinkles etched on our faces by age. But it is actually a virulent poison called botulinum toxin, and in my youth it was far more famous for causing fatal food poisoning. In those days, tinned corned beef was a staple food, but if the cans were not correctly sealed they could become infected with the bacterium *Clostridium botulinum*. The toxin the bacteria produced led to the death of anyone who unwittingly ate the contaminated corned beef.[1] Other meat products can also harbour the toxin: indeed, the word 'botulinum' derives from the Latin name for sausage (botulus).

Botulinum toxin is one of the most potent natural poisons known. An amount sufficient to cover the head of a pin is more than enough to kill an adult and it is estimated as little as a gram would kill a million people. It is alleged to have been used by the Czech resistance to assassinate the notorious SS-Obergruppenführer Reinhard Heydrich, a high-ranking Nazi whom Hitler considered as a possible successor to himself. Heydrich was attacked in Prague in the spring of 1942 by two British-trained Czech patriots, who lobbed a grenade, reputedly impregnated with botulinum toxin, into his car. Although Heydrich's injuries were not thought to be life-threatening, and the surgery he received was successful, he died eight days later from complications. Whether botulinum toxin actually caused his death remains highly controversial. Nevertheless, several countries have explored its potential as an assassination agent; the CIA, for example, laced some of Fidel Castro's favourite cigars with it (they were never used).

Botulinum toxin acts by preventing muscle contraction. When ingested, it gradually relaxes the respiratory muscles until they stop functioning, causing paralysis and death from asphyxiation. In the last decade or two, however, it was realized that if a minute quantity of the toxin is injected under the skin, it will paralyse the muscles in a highly localized way. At first this was used to treat people with an unfortunate condition in which their neck or shoulder muscles become permanently frozen, so that their head is twisted to one side. But it was soon recognized that botox had another effect – it ironed out the furrows produced by years of frowning and the crinkles due to smiling. As the Swiss physician Paracelsus once said, 'The right dose differentiates a poison and a remedy'.

The virtue of botox is that it binds so tightly to its target that it is only slowly washed away and the muscles remain relaxed for many months. But every six months or so, the procedure must be repeated. The downside is that the toxin also blocks contraction of the muscles used in facial expressions such as smiling, and thus tends to produce a smooth expressionless sphinx-like stare. Worse still, if too much is used it can cause local paralysis, resulting in droopy eyelids or a downturned mouth.

A Nobel Dream

Botox causes paralysis by blocking the transmission of electrical impulses from nerve to muscle. These two types of cell are not physically connected and the impulse cannot leap the gap that separates them. Instead, a chemical messenger is used to send signals from one cell to another. Transmission of information from nerve cell to muscle cell (or another nerve cell) takes place at specialized junctions called synapses, where the gap between the two cells is very tiny – less than one hundred-millionth of a metre (about thirty nanometres). For obvious reasons, the upstream cell that releases the transmitter (in this case the nerve cell) is known as the pre-synaptic cell and its target as the post-synaptic cell.

The tip of the nerve fibre is densely packed with small membrane-bound vesicles filled with a chemical transmitter. At the synapse between nerve and muscle the transmitter is acetylcholine, but many other chemicals are used to signal information between different types of nerve cells in the brain. When an electrical impulse arrives at the nerve ending it causes the vesicles to release their contents into the gap between the two cells. The transmitter that is liberated diffuses across the gap and attaches to a receptor on the surface of the post-synaptic cell, triggering an electrical impulse. In a muscle cell this electrical impulse causes contraction.

The use of chemicals to transmit information from nerve to muscle was discovered by Otto Loewi, a brilliant scientist who worked for much of his life in the Austrian city of Graz. Born in Frankfurt in 1873, he was initially more interested in painting, music and philosophy than in science and desired to become an art historian. Dutifully, however, he bowed to family pressure and went to medical school instead. There, he quickly became captivated by science. Loewi had an irrepressible zest for life, an enthusiasm for science and a sense of humour that lasted throughout his life: even during his last years, he maintained that 'excitement was good for him'.

In 1921, Loewi showed that chemicals are involved in transmission of electrical impulses between nerve and muscle cells. He attributed this breakthrough to an inspiration that came to him during a dream. He wrote, 'The night before Easter Sunday of that year I awoke, turned on the light, and jotted down a few notes on a tiny slip of thin paper. Then I fell asleep again. It occurred to me at six o'clock in the morning that during the night I had written down something most important, but I was unable to decipher the scrawl. The next night, at three o'clock, the idea returned. It was the design of an experiment to determine whether or not the hypothesis of chemical transmission that I had uttered seventeen years ago was correct. I got up immediately, went to the laboratory, and performed a simple experiment on a frog heart according to the nocturnal design.'

Loewi knew that if an electric shock were applied to the nerve innervating a frog's heart, the heart rate would slow. If, as he suspected, this was caused by a chemical transmitter released from the nerve endings, then the chemical should pass into the fluid bathing the heart. First in his dream, and then in reality, Loewi showed that if this solution were collected and then flowed over a second heart, the rate at which the second heart beat would also slow. Ergo, the first heart secreted a soluble chemical messenger that then acted upon the second.

Loewi called his chemical messenger Vagus-stoff because it was released by the vagus nerve that supplies the heart. Today, we know it to be acetylcholine, as indeed Loewi himself conjectured. Too cautious to publish a speculation that might prove premature, it was only after a considerable number of further experiments that he tentatively concluded, in 1926, that it might be acetylcholine. Loewi was lucky, because the nerve that supplies the frog's heart contains two sets of fibres: one set releases a chemical that causes slowing of the heart rate (acetylcholine) and the other secretes a substance (noradrenaline) that speeds it up. The frequency at which Loewi stimulated the nerve to the heart must have been exactly right to ensure that the effects of acetylcholine dominated. Once again, serendipity played a key role in scientific discovery.

In 1936, Loewi (together with his lifelong friend, the British scientist Sir Henry Dale) was awarded the Nobel Prize for his work. A mere two years later, on 12 March 1938, Germany invaded Austria. Loewi was informed late that afternoon that the Nazis had taken over the country, but being preoccupied with his research he ignored the significance of the news. He found out his mistake at three o'clock the following morning when a dozen armed storm troopers burst into his bedroom and dragged him off to prison, along with many other Jews. Uncertain of his future, and expecting to be murdered at any time, he scribbled his latest results on a postcard, addressed it to a scientific journal (*Die Naturwissenshaften*) and persuaded a prison guard to stick it in the post. He felt indescribable relief that his results would not be lost.

Two months later Loewi was released from jail and allowed to leave the country – but only after he had given all his assets to the Nazis, which included instructing a Swedish bank in Stockholm to transfer his Nobel Prize money to a Nazi-controlled bank as a ransom for his life. He escaped to England without a penny.

After spending time with Dale in London, Loewi held temporary appointments in Brussels and Oxford, before sailing to the United States in 1939 to take up a research professorship in pharmacology at New York University Medical School. He arrived in New York in June 1940, aged sixty-seven, armed with a visa and a doctor's certificate. Glancing at the latter while waiting to see the immigration officer he was horrified to discover it proclaimed 'Senility, not able to earn his living'. Fortunately the officer chose to ignore this questionable handicap, and Loewi was allowed into the United States. He was never bitter about the tumultuous upheaval to his life. In fact, he considered fate had been kind to him, as in the USA he was able to pursue his scientific endeavours at an age at which in Austria he would have been forced to retire. For another twenty-one years, he continued to inspire successive generations of students and spend his summers in ever-animated discussion at the Marine Biological Laboratories at Wood's Hole.

Hitler's Gift[2]

Sir Henry Dale was both a great scientist and a wise, influential and authoritative spokesman for science, who was held in affectionate veneration by all his colleagues. A tall, warm-hearted man, with a capacious memory, he played an important role behind the scenes in rescuing Jewish biologists, including Loewi, from Nazi Germany. He was also intimately involved in the discovery of chemical transmission. Like Loewi, Dale stressed the role of 'fortunate accidents' in his research. Equally important, however, was his perseverance. Dale's involvement in the story began in 1913 when he received an extract of ergot (a fungus) for routine testing. It had a potent and unexpected physiological action that aroused his interest. Using rigorous classical chemical methods, his colleague, the chemist Arthur Ewins, succeeded in identifying the active principle, which turned out to be acetylcholine. The physiological effects of acetylcholine mimicked those produced by stimulating certain nerves and led Dale to comment that if there was any evidence for the presence of acetylcholine in animal tissues it would be a good candidate for a neurotransmitter. World War I then intervened and Dale was occupied with other duties, but years later he succeeded in demonstrating that acetylcholine is indeed found naturally in animals and in isolating it from the spleen of a horse.

Dale's interest in acetylcholine was rekindled when he learnt of Loewi's dream-inspired experiment and he set out to see whether acetylcholine was secreted from nerve terminals at the nerve–muscle junction. This was not an experiment for the faint-hearted, for only tiny amounts of acetylcholine are released and, as Dale commented, it had an extraordinarily evanescent action. What Dale needed was a highly sensitive assay. Providence, in the form of the Nazis, provided it for him.

In 1933, shortly after coming to power, Hitler ordered that all Jews employed by state institutions should be sacked. Almost overnight, large numbers of academics were out of a job. William Beveridge, Director of the London School of Economics, inspired British academics not only to devise a rescue plan but to support it financially themselves. He also persuaded the Rockefeller Foundation to set up a special fund for Jewish scientists to enable them to move to American universities. Many Jews were helped to flee to the United States and Britain. They were Hitler's great gift to the Allies. Hitler was seemingly unaware of their value, reputedly stating that, 'If the dismissal of Jewish scientists means the annihilation of contemporary German science, then we shall do without science for a few years.' It proved to be a suicidal policy.

But it was to be of considerable value to Dale. Wilhelm Feldberg, who had developed a sensitive assay for acetylcholine, relates how one day in 1933 he was summarily dismissed from his post at the Physiological Institute in Berlin because he was a Jew. A few weeks later he sought help from a representative of the Rockefeller Foundation of New York who was visiting the city with the aim of trying to rescue the brightest Jewish scientists. While the man was sympathetic, he said, 'You must understand, Feldberg, so many famous scientists have been dismissed whom we must help that it would not be fair to raise any hope of finding a position for a young person like you. But at least let me take down your name. One never knows.' On learning Feldberg's name he hesitated, and riffling through his papers exclaimed with delight, 'Here it is. I have a message for you from Sir Henry Dale. [...] Sir Henry told me, if by chance I should meet Feldberg in Berlin, and if he has been dismissed, tell him I want him to come to London to work with me.'[3] Feldberg left Germany at once.

Feldberg's technique provided a highly sensitive bioassay that not only showed that acetylcholine was indeed released on stimulation of the nerve innervating a muscle, but also allowed the amount of the transmitter to be measured. This was achieved by flowing a solution over the nerve when it was stimulated, collecting the solution, applying it to a leech muscle and measuring the strength of the contraction it produced. The secret to the success of the leech test was the use of a chemical (eserine) that prevented the breakdown of acetylcholine by endogenous enzymes, thus prolonging its 'evanescent' action. It was the finding that acetylcholine was the transmitter at the nerve–muscle synapses which led Dale to share the Nobel Prize with Loewi in 1936. In his acceptance speech, Dale tentatively suggested that chemical transmission might not be confined to the neuromuscular junction but also take place in the central nervous system. His words were prophetic.

The War of Soups and Sparks

While Dale favoured chemical transmission, the flamboyant Australian neurophysiologist John Carew Eccles was equally certain that communication was electrical, believing that transmission at nerve–nerve synapses was too fast to be chemical. A long-running debate known as the 'war of the soups and sparks' began that greatly enlivened the previously rather staid scientific meetings of the Physiological Society. Young, excitable, domineering and extremely energetic, Eccles presented his views with characteristic forcefulness. Dale, a member of the Establishment, and by now both a Fellow of the Royal Society and a Nobel Laureate, adopted a calm magisterial air. Yet Dale and Eccles were essentially shadow boxing. Although their public debates could be tense, highly charged, and, to some observers, quite astonishingly adversarial, their personal relationship was far from acrimonious, as they exchanged friendly letters and shared their results prior to publication. Furthermore, their scientific disagreement provided a valuable incentive for them to seek far more evidence to support their ideas than might otherwise have been the case.

Eccles was impressed by the long time it took for the heart to slow down when the vagal nerve was stimulated. Since this was known from Loewi's work to involve a chemical transmitter, he inferred that the much faster transmission that took place at the junction between nerve and skeletal muscle could not be chemical and thus must necessarily be electrical. He was outraged at Dale and Feldberg's suggestion that acetylcholine mediated transmission at the nerve–muscle junction. By 1949, however, sufficient evidence had accumulated for Eccles to concede that transmission at the neuromuscular junction was indeed chemical.

He reserved judgement, however, about what happened at nerve–nerve synapses in the spinal cord and brain, remaining convinced that here, at least, electrical transmission might prevail. The crucial experiment that resolved the debate was carried out late one night in mid-August 1951 in Dunedin, New Zealand, by Eccles and his colleagues Jack Coombs and Lawrence Brock. Eccles claimed it was inspired by his conversations with the philosopher Karl Popper who argued that nothing could be proved in science, only disproved. So Eccles set out to prove that neurotransmission in the central nervous system was not electrical and – to his immense surprise –

he succeeded. The breakthrough came because the team used fine glass micropipettes, which they inserted into neurones of the spinal cord to pick up the electrical signals in the post-synaptic cells when the nerve was stimulated. Coombs, an electrical engineer, designed, built and operated the specialist apparatus needed to stimulate and record from the neurones. Eccles had arranged the experiment so that the potential trace would go down if transmission were chemical and would go up if it were electrical. It went down, and Eccles was momentarily stunned – the electrical transmission hypothesis was thereby falsified. It was a dramatic night in other ways too. Coombs's wife gave birth, her baby girl being delivered by his co-worker Brock (a medic) while Eccles continued to experiment into the early hours of the morning.

Thomas Huxley once described 'the slaying of a beautiful hypothesis by an ugly fact' as the great tragedy of science. But Eccles did not mourn the loss of his idea – he wrote immediately to Dale informing him he was now convinced that neurotransmission must be chemical. Dale replied, congratulating Eccles on the beauty of his observations and wryly commenting that 'your new-found enthusiasm is certainly not going to cause any of us embarrassment'. He later wrote that Eccles's conversion to the chemical hypothesis was like that of Saul on the road to Damascus, 'when the sudden light shone and the scales fell from his eyes'. It is one of the great strengths of the scientific method, and a measure of the quality of a scientist, that when the data show conclusively that a favoured hypothesis is wrong, it is quickly abandoned.

Mind the Gap

When the nerve impulse arrives at the end of the axon, it must somehow cause the release of transmitter from the tiny vesicles in which it is stored. Calcium ions play a crucial role in this process. The concentration of calcium ions is more than ten thousand times less inside our cells than outside it, and it is held at this level by molecular pumps that quickly remove any calcium that enters, either by ejecting it from the cell or by storing it in intracellular compartments. One reason calcium is kept so low is that it functions as an intracellular messenger, conveying information about events at the cell membrane to intracellular proteins and organelles. At the nerve terminal, for example, calcium triggers the synaptic vesicles to release the acetylcholine they contain into the gap between the nerve and the muscle.

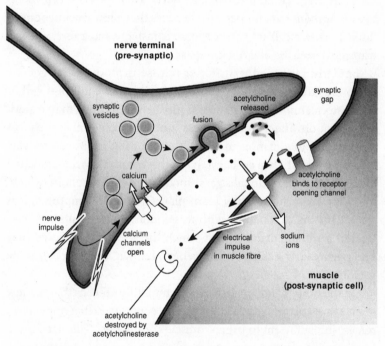

When a nerve impulse arrives at the nerve terminal, it causes calcium channels to open, allowing calcium ions to flood into the cell. This triggers synaptic vesicles filled with the neurotransmitter acetylcholine to move to, and fuse with, the cell membrane, releasing their contents into the synaptic gap. Acetylcholine then diffuses across the gap and binds to its receptors in the muscle fibre membrane. Binding of the neurotransmitter opens an intrinisc ion channel in the acetylcholine receptor, enabling sodium ions to enter the cell. The flow of sodium current triggers an electrical impulse in the muscle. In this way, the electrical signal passes from nerve to muscle via a chemical intermediary.

Calcium enters the cell via calcium channels in the pre-synaptic nerve membrane that open in response to the voltage change produced by the arrival of the nerve impulse. It is crucial that these channels open only when an impulse arrives and that they only remain open for a brief time; uncontrolled calcium influx can be dangerous as it triggers prolonged release of transmitter. One of the many toxic ingredients of the venom of the deadly black widow spider is alpha-latrotoxin, which inserts itself into the cell membrane, forming calcium-permeable pores that allow calcium ions to flood into the cell in an unregulated fashion and cause massive transmitter release and muscle spasms.

Similarly, increased transmitter release is produced by genetic mutations that prolong the duration of the nerve impulse and so increase calcium influx. People with such mutations may experience periodic attacks of dizziness, uncontrollable muscle shakes and uncoordinated movements so that they find it difficult to walk and lose their balance. They may also vomit. Attacks are often brought on by emotional stress, such as the excitement of watching your favourite football team play. Given the symptoms, it is not surprising that people with this condition are sometimes castigated for being drunk, a fact which the affected individual may find particularly galling if, as has been known, they happen to be teetotal.

On the other hand, inadequate calcium influx means that too few vesicles are encouraged to liberate their contents so that transmitter release is insufficient to trigger muscle contraction. This happens in Lambert Eaton myasthenic syndrome (LEMS), a condition in which the body produces antibodies against the calcium channels at the neuromuscular junction. These antibodies bind to the calcium channels and cause them to be removed from the nerve membrane so that nerve impulses fail to release any transmitter. The result is muscle weakness or paralysis. Most cases of LEMS are actually due to a tumour (usually a lung cancer) elsewhere in the body that possesses a similar type of calcium channel. The immune system's response to this cancer is to attack it by producing antibodies and those directed against the cancer cells' calcium channels cross-react with calcium channels present in the nerve terminals. LEMS is thus a red flag, warning the clinician to search for a possible tumour. It can be a valuable sign, for the sooner the lung cancer is treated the better the outcome for the patient.

All Docked Up and Ready to Go

It is sometimes envisaged that the interior of a cell resembles a pea soup, in which chemicals and organelles mill around in a random fashion. This is very far from the truth. Inside the cell everything has its place and is anchored in its correct position by a highly structured protein network called the cytoskeleton. This is particularly evident at the nerve terminal, where vesicles packed with transmitter are released only at specialized sites known as 'active zones'. Here several vesicles sit docked with the membrane, pretriggered for immediate release as soon as they get the signal to go. The calcium channels sit adjacent to the docking sites, reducing the distance that calcium has to travel once it enters the cell. This helps make transmission very fast. Within a millisecond (a thousandth of a second) of an electrical impulse reaching a nerve terminal, about 30 million molecules of acetylcholine are released. These quickly diffuse across the gap to their receptors on the muscle membrane, where they only remain bound for a couple of milliseconds, so that everything is over in about 20 milliseconds.

A small number of docked vesicles are 'trigger happy' and do not wait for a calcium signal: they are spontaneously released at a low frequency (they are too few to cause muscle contraction). Having a system that is all geared up and ready to go ensures that the arrival of a nerve impulse results in very rapid transmitter release – something you may be grateful for when your brain tells your hand to withdraw from a scalding-hot pan handle.

Complex molecular machinery is needed to overcome the enormous energy barrier that normally prevents the membrane of the synaptic vesicle fusing with that of the cell. This includes the numerous proteins that make up the docking and release complex, which act as molecular midwives, facilitating vesicle docking and membrane fusion. Precisely how binding of calcium ions to these proteins triggers the cascade of conformational changes that causes the vesicle and surface membranes to fuse is still unclear. However, inhibiting midwife protein function blocks neuromuscular transmission. Botulinum toxin, for example, prevents transmitter release and muscle contraction by destroying a specific set of these proteins.

Not all synaptic vesicles are primed for release. Most are stored at some distance from the release sites and must move to the membrane before they can be released; they must also undergo maturation processes that ready them for docking and release. Calcium also serves as a signal for mobilizing these troops of vesicles.

Poison Darts

Once liberated from the nerve terminal, acetylcholine diffuses across the tiny gap to the post-synaptic membrane of the muscle fibre, where it interacts with its receptors. Binding of the transmitter causes a conformational change in the acetylcholine receptor that opens an intrinsic ion channel, allowing a simultaneous influx of sodium and efflux of potassium ions. This decreases the voltage difference across the muscle membrane and (if it is sufficiently large) triggers an electrical impulse in the muscle fibre. In this way, acetylcholine serves to link the action potential in the nerve to one in the muscle, and ultimately to muscle contraction.

A large number of drugs and poisons work by interfering with the action of acetylcholine at its muscle receptor. The most famous is curare – the poison used by South American Indians to tip their arrows and the darts used in their blowpipes. Curare blocks binding of acetylcholine to its receptors in the muscle membrane and so prevents the nerve from stimulating the muscle fibre. Consequently, an animal hit by a dart is completely paralysed and falls out of the tree to the ground, where it is either slaughtered or dies from respiratory failure. Fortunately, curare is poorly absorbed by the digestive system, so animals killed in this way are safe to eat.

Curare was once also used in warfare and even the slightest nick with a poisoned arrow could be fatal. It was much feared. In his account of the discovery of Guiana (Guyana), Sir Walter Raleigh wrote that 'the party shot endureth the most insufferable torment in the world, and abideth a most ugly and lamentable death'. Depending on the dose, the victim can be awake, aware and sensitive to pain, but unable to move or breathe: unless given artificial respiration they will eventually die of respiratory failure. Toxins like curare have been used to tip arrows and spears for hundreds of years and the ancient Greek word toxicon, from which the name toxin derives, meant 'bow' or 'arrow poison'.

Curare can be extracted from many different South American plants, but the best known is the climbing pareira vine *Chondrodendron tomentosum*. The great Prussian explorer Alexander von Humboldt was the first European to describe how it is prepared, in 1800. He noted that the juice from the vine was extracted and mixed with a sticky preparation from another plant to make a thick treacly substance that could be glued to an arrowhead. Some years ago, I asked to view the curare-tipped blow darts owned by the Pitt Rivers Museum in Oxford. I was allowed to see them, but not to handle them or borrow them for a public lecture, for health and safety reasons. I was somewhat indignant, as I was certain the drug must have long since deteriorated. I was wrong: recently, curare that was 112 years old was shown to still be effective. Interestingly, the pure toxin was first isolated from a native preparation of curare held by the British Museum; it had been stored in a bamboo tube and the active alkaloid was hence named tubocurarine (tube-curare).

A disaffected group of conscientious objectors are reputed to have hatched a bizarre plot to kill the British Prime Minister Lloyd George using curare. Mrs Alice Wheeldon, her daughters Winnie and Hettie, and her son-in-law Alfred Mason all belonged to the No Conscription Fellowship that campaigned against compulsory military service (introduced because of very heavy losses on the Western Front) and the punishment and imprisonment of conscientious objectors (COs) during World War I. Alfred, a qualified chemist and a lecturer at Southampton University, obtained the curare. In late December 1916, the group was successfully infiltrated by two secret agents, Alex Gordon and Herbert Booth, who posed as conscientious objectors. Herbert Booth testified that he was given an airgun and pellets tipped with curare and detailed to shoot the Prime Minster as he walked on Walton Heath. There was enough curare to kill several individuals and, despite their protestations that the drug was intended to kill dogs guarding CO internment camps rather than the Prime Minister, and that this course of action had been suggested to them by Booth and Gordon, Alice, Winnie and Alfred were convicted of conspiracy to murder. Whether the assassination attempt could have been successful is unclear. Equally uncertain is if this was a genuine assassination attempt or a government plot to discredit the anti-conscription movement and the anti-war factions – were Booth and Gordon in fact agents provocateurs?

What is more definite is that the US government gave curare to operatives on covert operations, in case they were caught and had to take their own life to avoid being tortured. When Francis Gary Powers flew his U2 spy plane over Russia during the Cold War he carried with him an American silver dollar that had a small straight pin inserted in the side. The pin consisted of an outer sheath covering a fine sharp needle with a grooved tip coated with a brown sticky substance that Powers states he was told was curare. When he was shot down and captured, the Russians found the pin and tested it on a dog, which stopped breathing within a minute of being pricked and was dead thirty seconds later. It is questionable, however, whether curare was the sole ingredient on the pinhead as its action appears to have been unusually fast.

The poison hemlock (*Conium maculatum*) contains several alkaloids, but one of the most potent, coniine, acts like curare by blocking the action of acetylcholine and paralysing the respiratory muscles. The plant was used as a means of judicial execution in

Europe for centuries. Its most famous victim was Socrates, whose death is described in the *Phaedo*, which details how the paralysis developed, beginning with the feet and moving gradually upwards towards the chest.

Curare-like drugs, such as vecuronium, are often used in operations as muscle relaxants to enable the surgeon to operate more easily and to allow a lower level of anaesthesia to be used. This is especially important during abdominal surgery because contraction of the abdominal muscles might make it difficult for the surgeon to gain access without the intestines being squeezed out of the wound. Although the respiratory muscles are those least affected by curare, patients are usually artificially ventilated to help them breathe. The caveat with the use of curare-like drugs is that if anaesthesia is inadequate, the patient may be awake but unable to move, speak or communicate their distress. Each year, this happens to about 0.1 per cent of people undergoing surgery in the United States – approximately 25,000 people. About a third of them can feel the pain associated with their operation and the remainder have some awareness of what is going on without suffering pain. It is a particular problem during Caesarean sections when a lighter level of anaesthesia may be used to avoid anaesthetizing the unborn child.

Nerve Gas

Clearly, if a muscle is to be able to respond to a second nerve impulse, the first signal must be switched off rapidly. This is achieved in two ways. First, the transmitter remains attached to its receptor for only a short time before it spontaneously detaches. Secondly, the transmitter is rapidly removed. At the nerve–muscle junction, acetylcholine is destroyed within about five milliseconds of its release by an enzyme called acetylcholinesterase that sits in the synaptic gap.

Agents that inhibit the action of acetylcholinesterase are lethal. The most infamous is the nerve gas sarin. The Aum Shinrikyo sect came to public notice in 1995 when they released an impure form of sarin in the Tokyo Metro, killing twelve people, seriously injuring fifty and temporarily affecting the sight of almost a thousand more. The terrorist attack was timed to coincide with the morning rush hour.

Equally horrendous were the tests conducted forty years earlier by the British government. In May 1953, a number of young servicemen were asked to participate in trials for a new cure for the common cold. But the volunteers were misled in a brutal and unforgivable fashion, as they were not exposed to the cold virus but to sarin. Twenty-year-old Ronald Maddison died horribly forty-five minutes after the agent was dripped onto his skin, suffering from convulsions so severe it appeared to an eyewitness that he was being electrocuted. His lungs became clogged with mucous and he died of asphyxiation. Maddison was used as a human guinea pig to determine how much of the lethal agent was required to kill the enemy. His death was witnessed by a young ambulance man, Alfred Thornhill, who was traumatized by what he saw and afraid to speak out because the authorities threatened him with prison if he did so. The incident was quickly hushed up and only became known fifty years later when the Wiltshire police finally opened a second inquest into Maddison's death. The previous verdict of death by misadventure was overturned and replaced by one of unlawful killing. Maddison's sister said that until the inquest she and her family had never known the truth about how her brother died. Britain was not alone in wishing to test sarin on troops. Extraordinarily, in the 1960s, US military scientists requested permission from the Australian government to test the nerve agent on Australian troops.

Inhibition of acetylcholinesterase is fatal because it leads to a build-up of acetylcholine in the synaptic gap. The consequence is overstimulation of acetylcholine receptors, which results in muscle convulsions. Because acetylcholine is the transmitter at the nerves that innervate the glands, acetylcholinesterase inhibitors also cause excessive salivation, drooling and watering eyes. The symptoms of acute poisoning by sarin and other nerve gases are well described by the mnemonic SLUDGE: salivation, lacrimation, urination, diarrhoea, gastrointestinal upset and emesis (nausea and vomiting). The victim can also suffer from dizziness, skin irritation, tightness of the chest and involuntary muscle twitching. In the worst case, they may die by suffocation from convulsive spasms of the chest muscles.

Atropine is used to treat patients who have been poisoned by nerve agents. It acts by blocking acetylcholine receptors and thus reduces the ability of excess acetylcholine to exert its effect. Military personnel carry autoinjectable 'Combo' pens, consisting of a spring-loaded syringe containing a needle and a barrel filled with atropine which can be used to self-administer the drug rapidly in an emergency. The top of the pen also contains a Valium tablet to reduce levels of stress (which is probably much needed!). Too much atropine, however, can also incapacitate the soldier because it knocks out acetylcholine action too effectively. In this case, nerve–muscle transmission is prevented, resulting in muscle weakness.

Oximes are also used as antidotes to nerve gases, but are generally given in advance of a possible nerve agent attack. They reactivate acetylcholinesterase by removing the phosphate molecule that is added to the enzyme by the nerve agent.

The Deadly Calabar Bean

Another substance that inhibits the action of acetylcholinesterase is physostigmine, the active ingredient of the Calabar bean, *Physostigma venenosum*. The Nigerian name of the plant is esere, from which its alternative scientific name, eserine, is derived. It was eserine that enabled Feldberg and Dale to demonstrate that acetylcholine was released from nerve terminals, by preventing its breakdown by endogenous acetylcholinesterases.

The Calabar bean is native to Nigeria, where it has been used for centuries in tribal rituals to determine if a person is guilty of witchcraft or possessed by evil spirits. The accused is forced to swallow the chocolate-brown bean and is deemed guilty if they die, but innocent if they vomit up the beans. The outcome is actually dictated by the amount of poison the victim swallows, which depends on both the number and the ripeness of the beans, enabling those in power to manipulate the dose administered to achieve the verdict they desire. The Calabar bean is believed to have been used in a gruesome murder in which a young boy's headless and limbless torso was found floating in the Thames in September 2001. Analysis of his gut contents by the Royal Botanic Gardens at Kew revealed the remains of the Calabar bean plant, which both helped identify the country that the child came from and led detectives to speculate he was poisoned in a black magic ritual before being dismembered.

But agents like physostigmine also have therapeutic uses. Myasthenia gravis is an autoimmune disease in which the body produces antibodies against the muscle acetylcholine receptor. Each antibody bears two arms that grab two adjacent acetylcholine receptors and link them together, whereupon they are removed from the surface membrane and destroyed by the muscle cell. Consequently the number of acetylcholine receptors is markedly reduced, impairing nerve–muscle transmission and causing severe muscle weakness, progressive paralysis, and muscle wasting. Similar muscle weakness can be caused by loss-of-function mutations in the genes that encode muscle acetylcholine receptors. Children born with this disease have droopy eyelids, dropped jaws, open mouths and find it hard to stand, let alone walk. The treatment for both disorders is to increase the time that acetylcholine lingers at the synapse by inhibiting its breakdown by acetylcholinesterases.

The use of physostigmine to treat myasthenia gravis was pioneered by Dr Mary Walker, a quiet, unassuming assistant medical officer at St Alfege's Hospital in Greenwich. Noticing that the effects of myasthenia were similar to those of curare poisoning, she reasoned that they should be alleviated by physostigmine, a known antidote for curare poisoning. In 1934, she administered oral physostigmine to Dorothy Codling, a thirty-four-year-old chambermaid who had had the disease for six years. The effects were dramatic:

previously so weak she was unable to lift a cup and confined to bed, Dorothy was able to walk shortly after being injected with the drug. It became popularly known as the 'miracle of St Alfege's'. The treatment Mary Walker devised is still used today.

Riding the Lightning

Chemical transmission triumphed in the war of soups and sparks. Nevertheless, electrical transmission between cells does exist. At electrical synapses, the membranes of the communicating cells come extremely close to one another and are physically joined by specializations known as gap junctions. Each gap junction is made up of several hundred channels, packed tightly together in a semi-crystalline array. Uniquely, gap junction channels come in two halves, with one half of the channel being inserted into the membrane of one cell and the other half into that of its neighbour. When they couple up, a pathway is created for ions to flow directly from one cell to another, which allows electrical signals to spread quickly between cells.

Transmission at electrical synapses is about ten times as fast as at a chemical synapse because no time is needed for the transmitter to be released, diffuse across the synaptic gap and bind to post-synaptic receptors. As a consequence, electrical synapses often mediate defensive reactions such as the jet-propelled escape response of the squid, the ink cloud released by cuttlefish to cloak themselves from their enemies and the rapid withdrawal reflex of the earthworm that facilitates its backward retreat into its burrow when a blackbird pecks.

The rapidity of transmission also means that electrical synapses are perfect for synchronizing electrical activity in adjacent cells, and they are found throughout our bodies. Heart cells are wired together by gap junctions to ensure they contract in concert; gap junctions link the insulin-secreting beta-cells of the pancreas so that they release insulin simultaneously; and neurones in certain regions of our brains are electrically coupled so that they fire together. The pore of the gap junction channel is much larger than that of most other channels, which enables intracellular signalling molecules and small metabolites, as well as ions, to pass through. Consequently, gap junctions do not just connect cells electrically – they also ensure that the biochemical activities of adjacent cells are coupled. Gap junction channels even seem to play important roles in our skin, because inherited genetic defects that lead to their loss result in skin disorders. Those affected can develop thickened skin on the palms of their hands and the soles of their feet, as well as abnormalities of their teeth, hair and nails.

Leaping the Synaptic Gap

Although this chapter has focused on how the nerve impulse leaps the synaptic gap between nerve and muscle, synapses are not confined to the nerve–muscle junction. They are also found between nerve cells and gland cells and, very importantly, between different nerve cells, as described later. In all these places, the main mode of transmission is chemical and a cornucopia of different transmitters are involved. But why should chemical transmission be preferred over electrical?

One answer is that both its slower speed and the intricacies of its mechanism are better suited where integration of a plethora of signals might be advantageous. Another is that it may simply reflect the way in which cell signalling has evolved. Many simple organisms that consist of single cells, such as bacteria, communicate with one another via chemical messengers, enabling them to act as a vast team with coordinated defensive and attack strategies. Nor is the use of chemicals to transmit information from one cell to another confined to the nervous system. Long-range chemical messengers known as hormones transmit information between cells in our bodies that lie some distance apart from one another. Many different hormones circulate constantly throughout our bodies, influencing our mood, maintaining salt and water balance, stimulating cells to grow, readying our bodies to cope with stressful situations – even regulating the secretion of many other hormones. Pheromones wafted on the air enable communication between different organisms and act as sexual attractants, territorial markers and alarm signals. It seems likely that nerves have simply co-opted this universal chemical signalling system to serve their own ends.

5

Muscling in on the Action

Under a spreading chestnut tree
The village smithy stands;
The smith, a mighty man is he,
With large and sinewy hands;
And the muscles of his brawny arms
Are strong as iron bands.

Henry Wadsworth Longfellow, 'The Village Blacksmith'

Deep in the farmlands of rural Tennessee live some very unusual goats. Variously known as fainting goats, stiff-legged goats or myotonic goats, they topple over when startled. Their first name is misleading as the goats do not faint or even lose consciousness. Rather, they find it difficult to walk normally and often fall over when they are frightened because their muscles seize up and their legs become rigid. It resembles a kind of extreme cramp in which the muscles lock up so tightly that the animal can even be picked up without its legs bending – they are literally scared stiff. The attacks last only a few seconds and afterwards the goat is none the worse for its experience.

Mystery surrounds the origin of the myotonic goats. Anecdotal stories relate that in 1880 a man named Tinsley arrived at a farm in central Tennessee with a few goats and a zebu. He never said where he got the goats, or where he himself came from, and a year later, he moved on, leaving the goats behind. Another apocryphal story has it that their peculiar behaviour was discovered when one of them was shot for dinner and the rest of the herd collapsed in unison. What is certain is that a sudden stimulus such as a loud noise or an unexpected movement causes them to fall over – such as when a

Tennessee marching band goes past, or a passing train sounds its whistle. In some Tennessee towns a 'fainting' contest even forms part of an annual goat festival, with the prize being awarded to the goats that fall over the fastest and stay down the longest. Predictably, perhaps, animal rights protesters argue that scaring goats stiff is cruel. But myotonic goats are usually much-loved pets, kept for their novelty value rather than their meat.

Wiring our Muscles

The muscles that move our limbs are made up of many individual muscle cells, known as muscle fibres. These are grouped together into bundles, which gives meat its stringy texture. The nerve cells that control our muscles are known as motor neurones. If they are damaged or work imperfectly, our muscles are no longer able to respond when we wish to move them and gradually waste away from lack of use. This happens, for example, in motor neurone disease, where progressive degeneration of the motor neurones leads to weakness and muscle wasting, which results in a gradual inability to move the limbs and difficulties with speaking, swallowing and ultimately breathing.

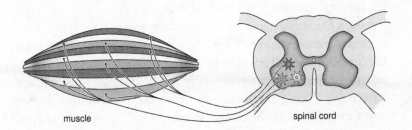

Three motor units, shown in dark grey, pale grey and white. Each motor nerve originates in the spinal cord and innervates many muscle fibres, located throughout the muscle.

Each muscle cell is innervated by a single motor nerve fibre, which has its cell body in the brain or spinal cord. However, one nerve cell can innervate several thousand muscle fibres because its terminal end splits into numerous branches. The nerve and its attendant muscle fibres are collectively called a motor unit and when the nerve fires all the muscle fibres it innervates will twitch in synchrony. The muscle fibres that make up the motor unit lie dispersed throughout the muscle, often quite distant from one another. Although this may seem strange, there is a good reason for it. It ensures that the force generated by stimulating a single motor nerve is spread throughout the whole muscle, and not focused in one place, which could cause the muscle to tear itself apart. In muscles that require fine control of movement, each motor unit is composed of a smaller number of muscle fibres: your finger muscles, for example, have fewer fibres per motor unit than your leg muscles.

The nerve contacts the muscle close to the centre of the fibre, where it splits into several fine branches, each of which forms a synapse with the muscle, as explained in the previous chapter. The muscle membrane opposite the nerve ending is thrown into numerous folds, which increases its surface area and enables many more acetylcholine receptors to be accommodated. Stimulation of the nerve releases a flood of acetylcholine, which diffuses across the synaptic gap and binds to these receptors.

Like a nerve fibre, and indeed all other cells in your body, muscle fibres have a voltage difference across their membranes, with the inside of the cell being more negative than the outside. Opening of the acetylcholine receptor channels dissipates this voltage difference, driving the membrane potential positive. Just as we saw with nerve cells, the change in membrane voltage opens muscle sodium channels and so sets off an electrical impulse (an action potential) that propagates along the muscle fibre in both directions from its point of origin. The action potential spreads rapidly over the surface of the muscle cell and then down into a network of tubular invaginations of the surface membrane, which penetrate right into the centre of the fibre. These conduct the action potential deep into the fibre interior, thus ensuring that all the contractile filaments contract in a single concerted step. The fact that an individual muscle cell contracts in an all-or-none fashion – completely or not all – was shown long before the all-or-none nature of the action potential was appreciated.

In a normal muscle fibre, a single nerve stimulus elicits a single muscle action potential that gives rise to a single contraction, such as when you blink your eye. It takes some time for the muscle to relax so that the duration of a muscle twitch is much longer than that of the electrical impulse. This means that if the muscle is stimulated repetitively, the twitches will summate to produce a sustained contraction of the muscle, known as a contracture. This enables you to apply a steady force to an object. The force a muscle can apply can be increased not only by stimulating an individual muscle fibre more frequently, but also by recruiting more motor units. Any sort of movement – from typing these words to hitting a squash ball – involves the complex coordination of a multitude of muscles and the precise control of their contraction by a myriad of electrical impulses in your nerves and muscles.

The muscle action potential is similar to that of nerve cells, in that it is initiated by the opening of sodium channels and terminated by opening of potassium channels. However, different genes code for the ion channels involved, which explains why a mutation in the muscle sodium channel does not affect nerve sodium channels (or vice versa) and why toxins that act on our nerves do not always affect our muscles. Muscle action potentials also involve more types of ion channels than those of axons. Of particular importance are the calcium and chloride channels, so-called because they are selectively permeable to these ions. Mutations in any of the different kinds of channel that shape the muscle action potential can cause muscle disorders.

Impressive: A Trojan Horse

Quarter horses were originally bred for quarter-mile racing (hence their name) and for handling cattle because they are very fast over short distances. Nowadays, they are more favoured as show horses. Some of the most beautiful have a mutation in their muscle sodium channel gene that causes a disorder known as hyperkalaemic periodic paralysis (or HYPP). Horses who carry the HYPP mutation are very sensitive to the concentration of potassium ions in their blood, and they become paralysed when this increases. Unfortunately, high concentrations of potassium are naturally present in alfalfa, so that eating hay made from alfalfa often causes attacks of flaccid paralysis. These start with muscle trembling and weakness, progress to swaying and staggering, and sometimes can be severe enough to cause the horse to collapse and fall over. Afflicted animals normally survive these attacks, but those that have the disease often have a shorter life span.

The mutation that causes HYPP prevents the muscle sodium channel from closing completely. Consequently, sodium ions continuously leak into the cell, decreasing the potential gradient across the muscle membrane and enhancing muscle excitability. This can cause the muscle to contract even when the horse is standing still. These spontaneous muscle contractions produce the peculiar impression of worms writhing just beneath the animal's skin. They also result in a striking muscular physique because the animal is, in effect, performing continuous isometric exercises. During an attack, the potential across the muscle membrane is reduced so much that the sodium channels shut down (they are said to be inactivated). Thus the muscle cannot sustain the contraction however much it is stimulated, the muscles become weak and floppy, and the horse falls to the ground.

Well-developed musculature is a highly desirable trait in a show horse, and animals with HYPP win many prizes. As a result of a programme of selective breeding for muscular physiques, 4 per cent of quarter horses are at risk of the disease. All of them can be traced to a single ancestor, a stallion called Impressive – named not for the quantity of offspring he sired, but rather for his powerful muscles. His majestic physique and the fact he always won his show classes meant he was in great demand as a stud. It was only later that it was appreciated that his offspring inherited more than his impressive muscles, for, like the original Trojan horse, Impressive's appearance concealed a more sinister gift.

Because only a single copy of the mutant gene is needed to cause the disease, roughly half of Impressive's offspring were susceptible to HYPP. Animals that carry two copies of the mutant gene are more severely affected. As a simple genetic test can provide a definitive diagnosis of HYPP, the disease could be easily eliminated if owners agreed not to breed from animals carrying a single copy of the mutant gene. However, this idea has proved somewhat difficult to enforce since affected horses win more prizes and are consequently more valuable. Nevertheless, since 2007, foals that carry two copies of the mutant gene are no longer eligible for registry with the American Quarter Horse Association.

Humans can also suffer from a similar condition. Those affected become weak and cannot move if they eat potassium-rich foods, like apricots and bananas, when they stop to rest after strenuous exercise or when they awake. During an attack their limbs become flaccid and floppy, like those of a rag doll. In a rare variant of this disease, known as paramyotonia congenita, people develop muscle stiffness when they get cold and it gets even more pronounced if they also exercise. The condition is not life-threatening, but it can be decidedly inconvenient to find that your hands become clamped to your spade when shovelling snow, that you cannot let go of your bicycle's metal handlebars when bicycling in cold weather, that you get stiff and weak when running in cold weather or that your jaw muscles stiffen up so much when you eat ice cream that you are unable to speak.

Many different mutations can cause HYPP in humans but, like the equine disease, they all make the muscle sodium channels leaky. In some cases, this causes the muscle fibre to become hyperexcitable, producing muscle trembling or stiffness, as in paramyotonia congenita. Other mutations render the muscle completely inexcitable so that it can no longer contract and the individual is paralysed. All of these conditions are precipitated by a small increase in the blood potassium concentration, which is why eating potassium-rich foods can trigger an attack. Although all of us will develop muscle weakness if the potassium levels in our blood increase too much, people with HYPP mutations are unusually sensitive.

Scared Stiff

Sodium channels are not the only muscle ion channels that can cause problems. An inherited disorder of muscle stiffness that resembles that of the Tennessee fainting goats is found in humans. It was first described in 1876 by the German physician Asmus Julius Thomsen, both in himself and his own family – over twenty members of his family, spanning five generations, were affected. Thomsen named the disease myotonia congenita, from myotonia, meaning 'muscle stiffness', and congenita, meaning 'inherited'. Interestingly, he hid his complaint until he was in his sixties, when one of his sons, who also suffered from the disorder, was accused of malingering and using his condition as an excuse to avoid military service. Thomsen published his studies to defend his son.

Characteristically, people with myotonia congenita are unable to relax their muscles easily; anything that involves a strong contraction tends to result in muscle 'cramp'. Pick up a heavy suitcase, for example, and they may find they are unable to release their grip when they put it down. One man who grabbed the door-pole of a moving tram, but slipped as he tried to leap on board, found he was unable to let go of the pole and was dragged along the road.

Muscle stiffness is worse when patients try to initiate a sudden movement after a period of rest, but gets gradually better with continued exercise. Even at rest, their muscles are continuously undergoing minute subthreshold contractions. Like horses with

HYPP, it is as if they are doing continuous isometric exercises. Consequently, they tend to have a fine physique and very well-developed muscles – so much so that they can resemble body builders. Yet despite an athletic-looking figure, their muscles often let them down. One patient recalls that having crouched down in the typical starting position for a race, his muscles would immediately freeze when he stood up. He said, 'My legs would lock into complete extension. It was like trying to run on stilts. I'd actually start running in maybe the last 20 or 30 metres.'

Goats Show the Way

The myotonic goats turned out to be the key to understanding myotonia congenita in humans. The Cincinnati physiologist Shirley Bryant had always been interested in muscle disorders. A brilliant and vivacious character who wore his hair in a ponytail even in his seventies, Bryant was a strong advocate of the idea that the study of animal diseases could elucidate similar human conditions, and he kept a strange menagerie of creatures, all of whom had inherited movement disorders, including tumbler pigeons, roller pigeons and a rare Australian marsupial, the Rottnest quokka. So when in the late 1950s he read of a herd of goats in Tennessee that fell down every time a train whistled past their field, he decided to investigate.

Bryant set off in a rented truck with an ex-convict driver. Initially, he had some difficulties in obtaining goats for his experiments. Fearing he would quickly complain that any fainting goats they sold him were defective and send them back, the farmers only gave him normal ones. Bryant had to make a second trip to Tennessee and it took some persuasion to convince the goatherds it was the stiff-legged animals he actually wanted.

Once back in his laboratory, Bryant removed a small piece of muscle from between the ribs of the goats under anaesthesia. The operation is simple, painless and does not hurt the animal: indeed, many patients have undergone similar biopsies. Importantly, these intercostal muscles are very short and provide a tendon-to-tendon preparation that ensures the muscle is undamaged. Bryant first inserted a fine glass electrode into the cell to record the potential across the muscle fibre membrane, and then a second electrode to stimulate electrical activity. He observed that in a normal muscle fibre the application of a small positive current produced a single action potential and a single muscle twitch. However, in muscle fibres isolated from myotonic goats the same current produced a burst of impulses that sometimes continued long after the stimulus had stopped. In effect, they were firing without being signalled to do so. This caused the muscle to enter an extended period of contraction and explained why the goats' legs became stiff and they fell over – their muscles were simply being excited more strongly. Similar findings were obtained using muscle biopsies from patients with myotonic congenita, showing that the human and goat diseases had a similar origin.

Unlike nerve fibres, muscle fibres have a high density of chloride channels, and in normal muscle the flow of chloride ions across the membrane dampens down electrical excitability, ensuring that a single nerve impulse produces only a single muscle twitch. Bryant conjectured that myotonic muscle might lack functional chloride channels, leading to enhanced excitability and sustained contraction. His experiments strongly supported this idea, although he was unable to measure the chloride current directly as there was no suitable voltage clamp for muscle at that time. Some years later, Richard Adrian,

normal muscle   myotonic muscle

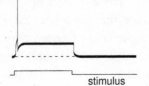
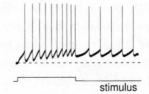

stimulus   stimulus

Myotonic muscle produces more electrical impulses when stimulated and these can continue even after the stimulus has stopped. If the electrical signal is played back through an audio amplifier, the normal muscle gives a single 'click' when stimulated, but the myotonic muscle sounds like an attacking dive-bomber.

a muscle specialist at Cambridge University in England, invented a new tool that was exactly what Bryant needed to confirm his theory.

In 1973, Bryant obtained permission to travel to England, together with four of his precious goats. It was not easy to take the goats to England because of fears that this might introduce a disease called bluetongue and a special Parliamentary Order, known as the 'Importation of Goats Order No.1', had to be passed to allow their importation. There was much media interest in the story, which was splashed across the front page of the *Wall Street Journal*. Despite this, the goats nearly did not make it. They were impounded at London's Heathrow airport because they did not have their papers. Bryant had left for Cambridge in advance in order to be there to meet the goats, leaving his colleague to complete their shipping papers – but his colleague forgot to do so! With the goats in imminent danger of being shot, Bryant telephoned his Cambridge colleague. Roused from breakfast, Adrian used his very considerable persuasive powers to convince the authorities that the requirement to immediately destroy animals lacking the necessary papers on arrival strictly applied only to those that were 'disembarking' – and not those that were 'disemplaning'. The goats were given a day's grace, by which time, fortunately, their papers had arrived.

A perfectionist, and something of a procrastinator, Bryant never got around to publishing much of the data he and Adrian collected in Cambridge using the voltage clamp. It joined a mass of other unpublished data that filled a whole filing cabinet (visitors to his lab were often amazed when he showed them elegant experiments, conducted years earlier, that he had never got around to publishing, but which shed light on some current problem). Nevertheless, their studies clearly showed that myotonic muscle has a lower chloride current and that this is sufficient to explain the repetitive action potential firing characteristic of myotonia.

In 1992, the gene that codes for the human muscle chloride channel was sequenced, enabling people with myotonia congenita to be tested for mutations in the gene. Almost immediately, the first mutations were identified – and shortly after they were found in the descendants of Thomsen. To date, tens of other mutations have been described in the chloride channel gene, including the one responsible for myotonia in goats, and all of them result in a loss of function. Moreover, it turns out that muscle stiffness, like that characteristic of myotonia congenita, can also be produced by mutations in other ion channel genes. Thomsen's myotonia, however, is of special scientific significance, as it was the first disorder to be linked to a defective ion channel. Many other ion channel diseases have been found subsequently, so many, in fact, that they have the distinction of having their own collective noun. They are known as the channelopathies.

Excitation–Contraction Coupling

The question of how the muscle action potential stimulates a skeletal muscle fibre to contract has occupied scientists for centuries. We now know that muscle contraction is triggered by an increase in the intracellular concentration of calcium ions. At rest, the calcium concentration inside the muscle cell is very low. Electrical stimulation of the muscle causes a dramatic rise in calcium, which binds to the contractile proteins and leads to shortening of the muscle.

However, the calcium ions do not come from outside the cell, but rather from a membrane-bound intracellular store called the sarcoplasmic reticulum. Calcium channels known as ryanodine receptors sit in the membrane of the sarcoplasmic reticulum and regulate the release of calcium. When they open, calcium floods out into the interior of the muscle fibre and triggers muscle contraction. When they close, calcium is quickly pumped back into the store, and the muscle relaxes. Ryanodine receptors are so called because they bind the plant alkaloid ryanodine with very high affinity.

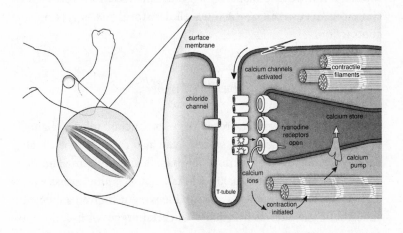

The membranes of our muscles are full of ion channels. Calcium channels in the tubular membranes sense the voltage difference across the surface and tubular membranes and pass this information onto the ryanodine receptors, which sit in the intracellular membranes of the sarcoplasmic reticulum, the muscle's calcium store. When the ryanodine receptors open, calcium rushes out, binds to the contractile filaments and causes the muscle to shorten. Muscle relaxation occurs when calcium is pumped back into the store, and its intracellular concentration falls. Chloride channels also sit in the surface and tubular membranes.

Precisely how the skeletal muscle action potential triggers the opening of the ryanodine receptors is still something of a mystery. After all, the action potential occurs in the surface membrane of the muscle and the ryanodine receptors are located in the membrane of the intracellular stores. Although these membranes are found in close apposition at specialized junctions, which lie within the tubular invaginations of the surface membrane, they never actually touch. It is clear, however, that voltage-sensitive calcium channels in the tubular membranes are somehow involved. One favoured idea is that the two kinds of calcium channel are in direct physical contact and that the ryanodine receptor channels, in effect, commandeer the tubular calcium channels' voltage sensor. As a consequence, a muscle action potential opens the ryanodine receptors, causing calcium ions to rush out of the intracellular stores and trigger muscle contraction.

Shiver My Timbers

Mutations in the ryanodine receptors – the calcium release channels of the intracellular calcium stores – can also cause trouble. Malignant hyperthermia is a rare disorder that affects only a small number of people (around 1 in 20,000 adults), but it is the anaesthetist's nightmare. It occurs when a susceptible patient is given a common anaesthetic gas, such as halothane, or certain types of muscle relaxant. These trigger spontaneous contractions of their skeletal muscles and a marked increase in muscle metabolism which increases heat production by the muscles and leads to a very rapid rise in body temperature – sometimes as much as 1°C every five minutes. In essence, the patients simply shiver themselves hot. An attack is a medical emergency as unless it is immediately treated, the increase in body temperature can be fatal. It is one of the leading causes of death from anaesthesia.

Malignant hyperthermia is also found in pigs, where it is known as porcine stress syndrome. It was once widespread in the UK pig population and was of considerable economic importance because not only do afflicted pigs often die, but the meat also becomes very pale, soft and unsaleable. As the name implies, the condition is triggered by various forms of stress including exercise, sex (in boars), parturition, being transported to market or simply by being kept in overcrowded conditions. It results from a mutation in the ryanodine receptor that causes the channels to become leaky so that the calcium concentration in the muscle cell rises, thereby stimulating metabolism, muscle contraction and an increase in body temperature. The skin of the animal becomes red and blotchy, and it may die of heat stoke within twenty minutes of the onset of an attack.

All affected pigs carry the same mutation and derive from a single founder animal in which the mutation arose spontaneously. The high incidence of porcine stress syndrome in the UK arose because pigs were selectively bred for lean meat and reduced back fat, attributes that unfortunately turned out to be associated with the gene for malignant hyperthermia. Lean, heavily muscled animals are much more likely to carry the gene. The porcine stress syndrome gene has now been almost completely bred out of the UK pig population by the simple expedient of giving each pig a whiff of a general anaesthetic (e.g. halothane) to breathe. Pigs that developed muscle rigidity and a rise in body temperature of $2°C$ within five minutes were removed from the breeding pool.

The pigs were the key to understanding the molecular basis of the human disease. Once the cause of porcine stress syndrome had been identified, corresponding mutations were soon found in the ryanodine receptors of about a third of families who suffer from malignant hyperthermia. It is thought that in affected people, anaesthesia causes the ryanodine receptor to become unusually leaky to calcium ions. These flood out of the intracellular stores, triggering sustained muscle contractions and muscle rigidity. In turn this stimulates muscle metabolism so much that the body temperature can soar to dangerous levels.

As the disease runs in families, it is now possible to test in advance if family members are likely to develop the problem when they are anaesthetized, and means can be taken to prevent it. In addition, the drug dantrolene sodium, which blocks calcium release from intracellular stores, is kept in all operating theatres for malignant hyperthermia emergencies. It was Shirley Bryant and Keith Ellis who first established how dantrolene acts. In doing so they saved many lives as the fatality rate due to an attack fell from 80 per cent in the 1970s to less than 10 per cent today.

Bryant had a longstanding interest in electricity, having started his career with quite a jolt. While training as an engineer he helped design an artificial lightning bolt for General Electric's exhibit at the World's Fair, and got a 30,000 volt shock for his pains. Fortunately for those with myotonia congenita and malignant hyperthermia, he survived.

As this chapter has shown, the electrical activity of our muscle fibres provides the trigger for muscle contraction, ensuring all parts of the fibre contract simultaneously; without it, our muscles simply would not work. In some animals, however, the muscle action potential has been commandeered for a very different purpose. Modified muscle fibres that have lost the ability to contract have evolved into specialized electric organs, in which the action potentials from many cells summate to produce a substantial electric shock. The next chapter, a slight but hopefully interesting digression, illustrates how animals use the currents produced by electric organs for such diverse purposes as offence, defence, navigation and communication.

6

Les Poissons Trembleurs

Who has not heard of the invincible skill of the dread torpedo and of the powers that win it its name?

Claudian, *Carmina minora*, XLIX (XLVI)

The electrifying discharge of the torpedo ray has been known since antiquity. It even features in the dialogues of Plato,[1] where Meno, perplexed by Socrates's arguments, compares the philosopher to the fish, saying, 'And if I may venture to make a joke, you seem to me both in your appearance and other respects to be very like that flat sea-fish, the torpedo. For this numbs those who come near it and touch it, as now you have numbed me, I think. For my mind and my tongue are paralysed, and I do not know how to answer you.' Other classical texts report that fishermen found that their hands were paralysed when they speared a torpedo or caught one in their nets. It is from this attribute that the fish derives its scientific name – the Latin word torpere means 'to be paralysed' – and it is from its Greek name – narke – that we get our word 'narcotic'. The puzzle for classical writers was that the numbing power of the fish could be experienced at a distance; it was not necessary to touch it.

Scientific investigation of electric fish began in the 1700s, when explorers returned from Africa with tales of a 'poisson trembleur' that made their muscles tremble intolerably when they touched it. This was the African catfish, *Malapterurus electricus*. The French naturalist Michel Adanson, who came across it when travelling in Senegal, was the first to compare the painful sensation with the jolt from a Leyden jar, and surmise that the fish could similarly deliver an electric shock.

Bas-relief from the tomb of Ti at Sakkura (*c*.2750 BC). The fourth fish from the left just below the boat, that lies beneath the pole, is the catfish *Malapterurus electricus*. The man sitting in the boat seems to be touching another fish with whiskers, probably also a catfish – if so, he is likely to be getting quite a shock.

The catfish was well known in ancient Egypt. It graces many tomb paintings and friezes, features in a bas-relief of a fishing scene from the tomb of Ti at Sakkura that dates back to as early as 2750 BC, and mummified catfish have even been found in the tombs of the Pharaohs. It also plays an important role in the myth of Osiris. Plutarch relates how Osiris was treacherously murdered by his brother Set and his corpse torn into fourteen pieces. His distraught wife Isis managed to retrieve all of her husband's dismembered body except for his penis, which had been thrown into the Nile and eaten by a catfish and two other fishes. As a consequence, perhaps, ancient Egyptians scrupulously avoided eating catfish.

Strangely, *Malapterurus* was considered as a love charm by Islamic authors and as an aphrodisiac by North Africans, despite an early missionary describing it as being 'of such a nature that no man can take it in his hand while it is alive, for it filleth the arm with paine as if every joint would go asunder'. It is no wonder it hurts, as the shock it delivers can be as much as 350 volts.

The strongest shocks of all are inflicted by the South American electric eel, *Electrophorus electricus*. Notwithstanding its common name, *Electrophorus* is not an eel but a member of the knifefish family: it simply resembles an eel. Jesuit missionaries were the first to describe it, in the sixteenth century, terming it the torpedo of the Indies. But it was not until the eighteenth century that people began to investigate its electrical nature and it was recognized that the paralysing effect of touching it was due to the fact it emitted an electric shock. Although some eels were eventually brought to the United States and to London, not everyone could afford to experiment on them as the price was as much as 50 guineas an eel, a considerable sum in those days.[2] Nor were the eels always in the best shape after their long journey. An alternative, and far more attractive, proposition for an intrepid young man was to go to the eel himself. One such was the scientist-explorer Alexander von Humboldt.

What a Stunner!

Spurred on by the desire for adventure and the wish 'to be transported from a boring daily life to a marvellous world', the twenty-nine-year-old von Humboldt sailed for South America on a journey of scientific discovery in 1799. His 'Personal Narrative' of the expedition, written on his return five years later, rapidly became a bestseller. Among others, it inspired the young Charles Darwin, who wrote that it 'stirred up in me a burning zeal to add even the most humble contribution to the noble structure of Natural Science'.

Von Humboldt was an accomplished experimenter with an avid interest in Galvani's work on frogs (which had been published a few years earlier). He was especially eager to obtain some electric eels, which were extremely common in the tributaries of the Orinoco River. But he found this far from easy because the fear of the eel's shock was so extreme that he was unable to persuade the local Indians to bring him any. Promises were forthcoming, but eels were not. Nor was money sufficient inducement, for it meant little to the local tribes. Frustrated by waiting, von Humboldt set out to catch them himself, so spurring his local Indian guides to offer to help by 'fishing with horses'. Von Humboldt wrote that 'it was hard to imagine this way of fishing; but soon we saw our guides returning from the savannah with a troop of wild horses and mules. There were about thirty of them, and they forced them into the water.'

He paints a vivid picture of the ensuing mêlée. 'The extraordinary noise made by the stamping of the horses made the fish jump out of the mud and attack. These livid, yellow eels, like great water snakes, swim on the water's surface and squeeze under the bellies of the horses and mules.' The horses, of course, endeavoured to escape, but they were prevented from doing so by the Indians, who screamed and yelled and prodded them back into the river with sharp-pointed sticks. The battle was intense. 'The eels, dazed by the noise, defended themselves with their electrical charges. For a while it seemed they might win. Several horses collapsed from the shocks received on their most vital organs, and drowned under the water. Others, panting, their manes erect, their eyes anguished, stood up and tried to escape the storm surprising them in the water.' Some finally made it to the bank, where they collapsed onto the sand, stunned and exhausted by the electric shocks.

Within just a few minutes the violence of the combat subsided and the battle was over. The exhausted eels drifted towards the bank and were easily caught with harpoons tied to long strings. Most of the horses survived. As von Humboldt acknowledged, those that died were unlikely to have been killed by the shock itself: they were simply stunned and then trampled underfoot by the other horses and drowned. This unique method of fishing was successful because, like an electric battery, the eel has a limited store of charge and its ability to produce electric shocks is rapidly exhausted. In the interval before it has recharged itself it can be captured without danger of electrocution.

Von Humboldt's interest in *Electrophorus* extended beyond the scientific. He also noted that its flesh did not taste too bad, although most of the body was filled with the electrical apparatus, 'which is slimy and disagreeable' to eat.

A Shocking Use of Muscle Power

The electric eel can produce a powerful shock of over 500 volts and a current of one ampere, which amounts to a power output of 500 watts.[3] This is sufficient to run several lightbulbs, as one Japanese aquarium demonstrated when it wired up an electric eel to power its Christmas tree lights. It is also enough to stun, or even kill, a human or large animal. In von Humboldt's time so many mules were slain at the ford across one river that the road had to be redirected, and even in the mid-twentieth century ranchers were losing (or thought they were losing) cattle to eels in such large numbers that they instituted 'electric eel drives' in which the fish were encouraged to shock themselves into exhaustion and then were killed using machetes with insulated handles.

The physiological effect of a shock from an electric eel is no different from that produced by an artificial electric current of similar magnitude. It can cause involuntary muscle contraction, paralysis of the respiratory muscles, heart failure and even death, either by electrocution or more often by drowning as a result of the victim being stunned. It can also be very painful. Von Humboldt once inadvertently stepped on a large excited eel that had just been taken from the water almost fully charged. He described the pain and numbness as extreme, complaining that 'All day I felt strong pain in my knees and in all my joints', accompanied by twitching of the tendons and muscles (hence the Spanish name of the fish, tembladores). It is perhaps not surprising that the Indians of the llanos feared them.

Electric eels have no teeth and must swallow their prey in one gulp, which is obviously harder if it is wriggling and may be why they generate electric shocks to stun their prey. Much of the time they lurk in the mud on the river bottom, but as they get most of their oxygen by gulping air they must surface every few minutes or so to breathe. Because they breathe air they do not die if they are removed from the water and so can be easily studied. I vividly remember visiting a lab that worked on electric eels many years ago and being taken to see the fish. Before entering the room I was required to don rubber gloves that came up to my armpits, in case a fish leapt out of its tank and I inadvertently came into contact with it. It made quite an impression.

 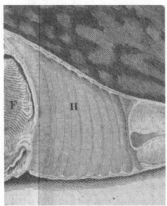

Left. Volta's electric battery, consisting of stacks of silver (A) and zinc (Z) discs. *Right.* Cross-section through the body of the torpedo ray showing the columns of electroplaques (H indicates one stack). The resemblance is remarkable.

Electrophorus has a long, cylindrical, eel-like body, with a dark-grey back and yellowish belly, and it can reach an enormous size. Larger specimens weigh over twenty kilograms, exceed two and a half metres in length and are as thick as a man's thigh. The vital organs are crammed into the front one-fifth of the body: the rest of the fish houses the backbone and swimming muscles, but most is pure power pack. The main electric organs lie on either side of the eel's body. Each contains thousands of modified muscle cells, known as electroplaques, which have lost the capacity to contract and are specialized for producing an electric discharge. These wafer-thin, plate-like cells are stacked up in long columns, like a giant pile of coins, with as many as 5,000 to 10,000 cells per column. There are around seventy such columns on each side of the eel's body. Each stack of electroplaques bears a strong similarity to a voltaic pile – the primitive battery discussed in Chapter 1 – a fact which Volta himself noted.

Throwing the Switch

The two faces of the electroplaque cell are markedly different. One side is smooth and criss-crossed by many nerve endings: the other is deeply invaginated and is not innervated. At rest there is no difference in voltage between the two outer faces of the cell and thus no shock is produced. When the fish decides to zap its prey, it fires off an impulse down the nerve supplying the electric organ. This triggers an electrical impulse in the electroplaque – in effect a muscle action potential – that is confined to the innervated side. As a consequence, a voltage difference develops across the two sides of the cell of as much as 150 millivolts. Because this happens simultaneously in all electroplaques, and because they are arranged in series, the voltages add up to produce a considerable jolt of 500 volts or more (about four times as much as a household electrical socket in the USA and twice as much as one in Europe). Thousands of muscle action potentials, all firing in synchrony, thus underlie the shock.

In essence, each electroplaque behaves like a miniature living battery with the stimulated side (facing the tail) bearing a negative charge and the opposite side (facing the head) a positive charge. These tiny batteries are stacked up in a head-to-tail fashion in a long column. A simple analogy is with a battery-powered torch in which the cylindrical handle contains several batteries, stacked one on top of another (positive to negative). Their individual voltages add up to give the total needed to power the torch. In the same way, the tiny voltages produced by the individual electroplaques when they are excited add up to give a very large voltage. The more cells in the stack, the bigger the jolt. Young eels, which have fewer electric cells per stack, can still produce a significant shock, but it is much less than their full-grown cousins. Each individual shock does not last long, as the electrical impulse at the innervated face of the electroplaque is all over within a couple of milliseconds. However, the eel produces a barrage of jolts by firing off rapid bursts of impulses in quick succession – as many as 400 a second.

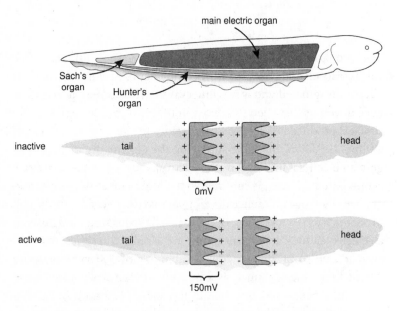

Top. The electric eel has three electric organs, but only the main one generates the large electric shock it uses to stun its prey. *Middle and bottom.* Two of the wafer-like electroplaques that make up one of the columns of the main organ. When a cell is at rest (inactive), its inside is negatively charged and both its external faces are positively charged; thus there is no voltage difference between the two outside faces. When the eel fires a shock (active), the voltage at the posterior face of the electroplaque becomes negatively charged, so that the voltage between the two outside faces of the cell now amounts to around 150 millivolts. The voltages from the individual electroplaques summate to deliver a substantial shock

Although the voltage difference between one end of a stack and the other is considerable, the current that flows out of the end of the stack into the surrounding water is relatively small. This is advantageous, as it is not enough to fry the eels' own cells. However, the currents through the whole collection of parallel stacks add up, so that the total current generated is much more – it amounts to about an amp. The space between each electroplaque is filled with a highly conductive jelly-like material, which is probably what von Humboldt found so distasteful to eat. This serves a very important function; it ensures that the current flows easily from one electroplaque to the next in the stack, and between the end of the column and the surrounding water. Equally important is that each column is well insulated along its length, in order to coax the current to flow along the column rather than leaking out sideways into the eel's own surrounding tissues.

It is clearly valuable to have the electroplaques as thin as possible, because the more cells that can be crammed into the column, the greater the voltage that can be developed, and the larger the shock produced. However, the thinner the cell, the more quickly it fills up with sodium ions, which enter during the electrical impulse. This creates problems because it reduces the concentration gradient that drives sodium ions to move into the cell, which means that during a train of impulses the size of the electrical impulse each cell produces steadily falls. Consequently, the magnitude of the shock, and the frequency at which it can be generated, gradually decreases and finally fails. The electric organ is then discharged – just like an overworked battery. It was this phenomenon the Indians exploited in their novel fishing technique. Recharging the electric organ takes some time and is achieved by molecular pumps that laboriously pump all the sodium ions that have entered the cell back out again, thereby restoring the sodium gradient that powers the electrical impulse.

Zapped!

The electric ray *Torpedo* uses a system similar to that of the electric eel to produce an electric shock, but with some modifications because it is a marine fish rather than a freshwater one. In freshwater, there are few dissolved salts to carry an electric current, so it does not travel very far and the eel must be close to its prey to order to stun it. Thus the eel generates a much greater voltage, which helps force the current through the water. Seawater is a far better conductor of electricity than freshwater because it contains more salts, so the magnitude of the current diminishes less rapidly with distance. The torpedo is perfectly adapted to its marine environment as it generates a higher current but a lower voltage than *Electrophorus*.

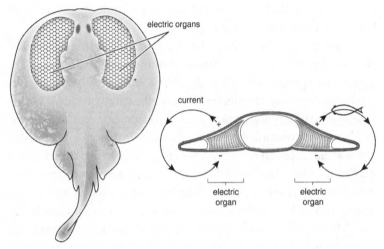

The electric organs of *Torpedo* lie on either side of the head. The path of current flow when the electric organs discharge is shown in the cross-section through the fish on the right.

The torpedo has two large kidney-shaped electric organs positioned one on either side of the head. Each consists of 500 to 1,000 closely packed stacks of electroplaques, and there are around 1,000 cells in each column. Because there are fewer cells per stack, *Torpedo* cannot generate as high a voltage as *Electrophorus*; the maximum shock is only about fifty volts, around a tenth that of the eel. However, the current is greater because of the much larger number of columns, so that the torpedo can produce a current as high as fifty amps and a power output of more than a kilowatt at the peak of its discharge. The fact the torpedo generates more amps and fewer volts than the electric eel is dictated by the greater conductivity of the medium in which it lives. The exigencies of marine life also explain why its electric organs are short and wide whereas those of the eel are long and thin: this is because you need many short stacks to get high current with lower volts.

The columns of electroplaques are stacked vertically between the upper and lower surfaces of the wings of the ray. When the electric organ discharges, the current spreads out into the surrounding medium, being greatest directly above or below the electric organ. The hunting behaviour of the torpedo exploits this fact. It rests on the bottom of the sea floor until a fish comes close, whereupon it swims upward, emitting a stunning series of electric shocks and orientating itself so its prey will receive the greatest jolt. It then drops down onto the immobilized prey, wraps its wings around it and manipulates it into its mouth.

As in *Electrophorus*, only the lower surface of each of the torpedo's electroplaques is innervated. This modified muscle membrane is packed with so many acetylcholine receptors that they form a semi-crystalline array. In essence, it is one giant synapse. Excitation of the nerve supplying the electric organ releases the transmitter acetylcholine (see Chapter 4), which opens acetylcholine receptors in the electroplaque's lower membrane and produces a potential difference between one side of the cell and the other of around 100 millivolts. This is significantly less than that produced in the electroplaques of the electric eel. Nevertheless, the main reason the torpedo generates a lower voltage is because it has fewer cells per stack. It takes a lot of energy to produce an electric shock and it cannot be maintained continuously so, like the electric eel, the torpedo produces bursts of pulses (about 100 per second), with each shock lasting just a few milliseconds.

Why Does the Torpedo Not Shock Itself?

Why the torpedo (or indeed the electric eel) is not incapacitated by the shock it produces is a puzzle that is still not fully understood. Current flows from one end of the electroplaque stack to the other and then out through the tissue and skin into the water. Because the electric organs sit in the wings, the current does not flow directly through the torpedo's heart or brain. Furthermore, the current flowing through any individual part of the fish is small as each column of electroplaques produces only a small amount. The prey, however, experiences a substantial shock because the weak currents through the different columns add up to create a much greater current in the water. It is also believed that fatty layers in the fish's skin act as an insulator to protect it from its own shocks, because if the skin is scratched or damaged (which renders this insulation less effective) an electric eel twitches when it discharges, suggesting it now feels the shock. Of course, it is also important that the skin above the electric organs is not well insulated so that the current can escape into the water and, as expected, the skin over the top and bottom of the torpedo's electric organs is of higher conductance than that covering other areas of the body.

Shark Attack!

In September 1985, the telecommunications company AT&T laid an undersea fibre optic cable between Gran Canaria and Tenerife in the Canary Isles. A mere month later, the cable shorted out ten kilometres (six miles) out from Tenerife at a depth of 1,000 metres, interrupting telecommunications. AT&T was faced with the laborious, time-consuming and expensive task of raising the cable and replacing the damaged section. Mysteriously, the cable developed a similar fault twice the next year and yet again in April 1987. Careful examination of the damaged cables revealed that they were studded with sharks' teeth, suggesting that the damage was caused by a shark's bite. The main culprit was the crocodile shark, *Pseudocarcharias kamoharai*, which has very powerful jaws.

To understand what was happening, AT&T went fishing. Hundreds of sharks were caught and examined. In a bizarre experiment they even tried force-feeding one shark a sample of cable. 'He was not happy about having someone try to shove it down his mouth,' Mr Barrett of AT&T reported.

Fibre optic cables are supplied with undersea repeater stations along their length that boost the optical signals. The high voltage required to power these amplifiers is supplied by a copper sheath surrounding the optical fibre core, and what seems to have happened is that the shark bit through the insulation, exposing the copper sheath to seawater. This short-circuited the power system and thereby interrupted communications.

Remotely operated vehicles have filmed sharks biting electric cables, and one shark was even observed to come back for a second bite when the cable slipped from its mouth. The problem with fibre optic cables is that they are much thinner than the old-fashioned copper wire variety – often only the diameter of a garden hose (around one inch) – and thus much more vulnerable to a shark bite. Furthermore, the shark does not need to sever the cable to cause significant damage – a sharp kink is sufficient. Eventually AT&T solved the 'Jaws problem' by encasing the cable in two layers of steel tape and a thick polyurethane coating. They also discovered that sharks normally do not feed below about 2,000 metres, so extra protection from shark attacks is not required in deeper waters.

Electrosensory Perception

But why did the shark attack the cable? The high voltage the cable carries generates electrical and magnetic fields around and along the cable. It is presumed that the sharks were attracted by the surrounding electric field, as a shark can sense the tiny electric field caused by normal muscle activity in other organisms and so detect its prey even if it is well camouflaged. Even when access to olfactory clues is prevented, a hungry shark will find a flatfish buried in the sand. It will also become excited and 'attack' an artificial electric field of a similar magnitude to that generated by the breathing movements of the flatfish. A mere four microamps of current is sufficient, so it is not surprising that stray signals from underwater cables can be detected.

All organisms generate tiny electric currents when their nerves fire impulses or their muscles contract. It is not enough to stay still – breathing movements or a beating heart will give you away. As you read this, the muscles in your body are producing a background electrical crackle. Fish that live in seawater are highly sensitive to such stray electric currents because the resistance of the water is low (due to the salt it contains), so the current travels further: some fish can detect fields as small as 0.01 microvolt per centimetre (one ten-thousandth of an AA battery). A human standing still and immersed in seawater up to their neck will produce an electric field of about 0.02 microvolts per centimetre for about one metre around their body, which is easily large enough to be sensed by a shark.

Sharks are not alone in being able to detect an electric field – many other fish are able to do so, including catfish, rays, lampreys, lungfish and coelacanths. It is thought that some can even detect the change in the Earth's electric field that precedes an earthquake. This may be the origin of the Japanese legend that earthquakes are caused by a giant catfish, known as namazu. This fish features in numerous beautiful ukioy-e woodblock prints and, more prosaically, on modern Japanese earthquake early warning devices.

'Electroreception' has evolved independently on more than one occasion because the sense organs used to detect electric currents differ between different types of fish. The electroreceptor cells that give the sharks and rays their exquisite sensitivity to electric fields lie in specialized sense organs known as the ampullae of Lorenzini,[4] which are concentrated in the head of the shark, around its

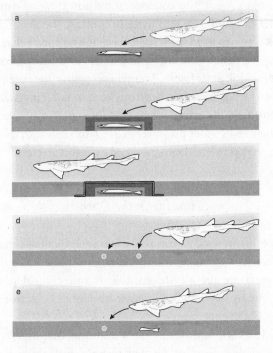

Adrianus Kalmijn's classic experiment that showed how sharks use electroreception to locate their prey. The sharks were collected from the English Channel and the North Sea and studied in captivity. (a) A flatfish introduced into a tank immediately buries itself beneath the sand but is instantly found by a hungry shark. (b) The shark still finds its prey even when the flatfish is placed in an agar chamber and covered with sand to prevent visual, mechanical or chemical clues giving away its location. As the agar has the same conductivity as seawater, it does not act as a barrier to electric signals. (c) The feeding response is abolished by covering the agar chamber with a thin film of plastic, whose electrical resistance is high enough to block out the electric field of the flatfish. This suggests the shark may be detecting the tiny electric currents produced by muscles of the flatfish as it breathes. (d) Crucially, when the flatfish is removed and replaced by a pair of electrodes that emit an electrical signal similar to that of the flatfish, the shark attacks the electrodes and tries to eat them. (e) In fact, it is even more interested in the electrodes than a piece of fish, indicating that at short range the electric field is a much stronger directive than visual or chemical stimuli.

nose and mouth. We still do not understand how these cells manage to achieve their extraordinary sensitivity. In contrast, the electroreceptors of bony fish are modified from the lateral line receptors that are used to detect movement. Next time you prepare a whole fish for dinner, take a careful look at its flanks. You will see a fine line running along the centre of its side that stretches from head to tail. This is the 'lateral line'. In most fish, the sense organs that form part of the lateral line perceive changes in water pressure. In a few fish species, however, the lateral line receptors have been modified to detect an electric field.

Hunting in the Dark

Some amphibians, such as the axolotl and the giant salamander, as well as primitive egg-laying mammals (monotremes) like the duck-billed platypus, are also electrosensitive. It is no coincidence that all of these animals live in aquatic environments as electroreception needs a conductive medium.

The platypus is a most remarkable mammal that lives in streams in Australia. It is covered in fur, has webbed feet, spurs filled with poison on its hind legs, a flexible, rubbery beak shaped like that of a duck, and it lays eggs. It also has a highly developed electroreceptive sense that enables it to catch prey in muddy streams at night even though it closes its eyes, ears and nostrils during a dive. The skin of the bill is rich in electrosensory cells, as many as 40,000 of them, that are arranged in long rows that run from the base of the bill to its tip. The electroreceptive system is highly directional, and as it hunts the platypus sweeps its head from side to side. This may help it locate its prey by enabling it to compare inputs from electroreceptors on the left and right side of the bill, in the same way that you swing your head from side to side when trying to locate a sound source. Unusually, the platypus is also able to determine the distance to its prey. It does so by integrating its electrical and mechanical senses, using the delay between the arrival of electrical signals and pressure changes in the water produced by movements of the prey to gauge distance.

The Western echidna or spiny anteater, a terrestrial monotreme, has a similar but less complex electrosensory system. It looks a bit like a long-nosed hedgehog and uses its snout to probe damp leaf litter for the earthworms and other invertebrates on which it feeds. Electroreceptors concentrated in the skin covering the tip of the snout help detect its prey. The short-beaked echidna, which has far fewer electroreceptors, feeds on ants. It is thought it may only use its electrosense after rain, when it feeds particularly actively.

The electroreceptors in monotremes are quite different from those of fish and appear to have evolved from mucous glands. This is helpful for an animal that is not fully aquatic as it ensures the sensory cells remain moist, increasing their ability to detect an electrical signal. It is the bare nerve endings that serve as electrodetectors – there is no specialized sense organ. Although an individual sensory nerve fibre ending has a detection threshold of just one to two millivolts per centimetre, the platypus can detect field strengths almost a hundredfold smaller. The remarkable sensitivity of the platypus probably originates from its ability to integrate information from many thousands of receptors, thus markedly increasing its ability to detect a signal.

The Guiana dolphin lives in the coastal regions and estuaries of the north-east coast of South America, where suspended silt and sediment can cloud the water. It uses electrosensors located in pits in its 'beak' to detect the weak electric fields emitted by small fish. It seems likely its electrosense functions as a supplementary means of detecting prey at close quarters.

Finding One's Way

The shocks produced by the electric eel were a concern to Charles Darwin, who found it difficult to see how they could have evolved since the organ was of no defensive or offensive use until fully formed. What possible advantage could the ability to produce a small electric shock be to an animal, he wondered. But, as we now know, a weak electric discharge is actually of considerable value.

Fish that produce weak electric pulses, just a few volts in magnitude, were discovered in the late nineteenth and early twentieth century. These fish have a sophisticated electrosensory system that combines generation of weak electric shocks with electroperception. It is used for detection of predators and prey and is also invaluable for finding their way around in the murky waters in which they live, where it is too dark to see. Passive electroreception such as that of the shark is like hearing: it simply detects electric fields in the environment. Active electroreception is more like radar: the fish produces an electric field and detects objects by the way in which they distort the field.

The crucial experiments that showed the function of these weak electric discharges were carried out by Hans Lissmann and Ken Machin in the 1950s. Lissman was fascinated by his discovery that the knifefish *Gymnarchus* frequently swam backwards without bumping into anything, that it could navigate easily around obstacles, and that it could locate its prey from some distance away, despite its vision being very poor. There is a story, possibly apocryphal, that the ability of *Gymnarchus* to detect an electric current was revealed when a student combed her hair near its tank and the fish promptly went wild. This may be a myth, but Lissman did report that combing his own hair had this effect (the electrostatic charge generated may have stimulated the fish). By placing electrodes in the fish tank, he found that the fish generated a constant stream of electrical pulses and that it was very sensitive to any changes in the electric field that it set up. His paper concludes forlornly on a poignant note: 'Unfortunately, while the investigation was still in progress my *Gymnarchus* died, and it appears very difficult to replace this animal [. . .] I should be grateful if anyone who could suggest a possible source of supply would write to me.'

Apparently no one did, because shortly after, in 1951, Lissman set off to Africa to obtain some more specimens. His destination was the Black Volta River in the northern territories of Ghana.

During the rainy season the river is extremely muddy as it contains very high levels of suspended particulates. Thus not only is it impossible for the fish to see their prey, it was also difficult for the scientists to see the fish. They detected their presence using a pair of electrodes that they suspended from a long pole on the riverbank (or the boat) and wired up to an amplifier that converted the electrical signal into an audible one. It is possible to 'hear' electric fish in this way, and Lissman commonly picked up a uniform high frequency hum of around 300 cycles per second. This enabled him to capture several fish, three of which he succeeded in bringing back alive to Cambridge, where he was able to study them in more detail.

Lissman and Machin set out to test the idea that *Gymnarchus* could detect objects in the water by sensing the distortions they cause in the electric field the fish itself produces. They used porous earthenware pots of different electrical conductivity: some pots were filled with distilled water so that they had low conductivity, whereas others were filled with a concentrated salt solution to simulate the higher conductivity expected of a fish. What they found was that *Gymnarchus* was easily able to discriminate between pots of different conductivity.

The electrosensory apparatus of *Gymnarchus* consists of an electric organ that generates a weak electric field and a detector system that picks up distortions in this field produced by objects in the environment. In effect, the fish is creating an electrical image of their environment analogous to the visual one we use to navigate. The electrical impulses such fish generate are relatively weak – less than a volt. They are produced by an electric organ which works in a similar way to that of the electric eel, but because there are fewer electroplaques the charge they produce is correspondingly smaller. The electric field generated by an electric fish looks somewhat similar to the pattern of iron filings placed around a bar magnet. Lines of force (at the same potential) run from the head to the tail, becoming increasingly weaker the further away they are from the fish. The current flows at right angles to the isopotential lines and thus leaves the fish at right angles to the body and enters it at the tail.

If an object is placed in this electric field it will distort it. For example, if its resistance is greater than that of the water (as in the case of a rock) electrical current will be shunted around the object, causing a local decrease in current density and an 'electrical shadow' on the surface of the fish. On the other hand, if the object has a lower resistance – another fish, for example – then the current will preferentially flow through it increasing the current density and creating an 'electrical spotlight' on the skin. The closer the object, the larger the spot. By detecting these changes in current strength the fish can not only sense the presence and size of the object but also what it is made of, enabling it to decide whether to attack, run away from, or simply ignore it. Of course, if the object is exactly the same resistance as the water, it is not detected.

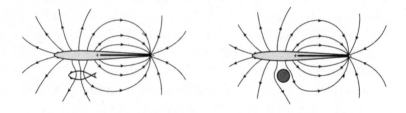

The electric field around *Gymnarchus* is distorted both by an object of higher conductivity than the water, like a fish (*left*), or of lower conductivity, such as a rock (*right*). The lines represent the flow of electric current.

Electroreceptors in the skin of the fish monitor its own electric field and the distortions produced by objects in the environment. In knifefish like *Gymnarchus*, there are about 15,000 of them, concentrated in the head, but also found at lesser density along the top of its back. There is an especially sensitive 'hot spot' of receptors located on the jaw. These tuberous electroreceptor organs consist of a small pit, lined at the bottom with sensory cells that act like miniature voltmeters and detect the voltage drop across the skin. They are extremely sensitive: when Machin built an electrical model to try to simulate the fish's electrical sense the fish triumphed over it every time.

Speaking in Sparks

The discharges produced by electric fish can be grouped into two kinds: pulse-type and wave-type. Pulse-type electric fish like the elephant nose fish *Gnathonemus* emit a stream of brief electric pulses a few millivolts in amplitude. Wave-type electric fish such as the knifefish *Gymnarchus* emit a continuous electric current that oscillates in strength. These sinusoidal oscillations are extremely constant – almost as good as a commercial oscillator – and are produced at a frequency of 800 to 1,000 cycles per second.

Both types of fish are able to tune the frequency of their signals, which can vary not only between species and sexes, but also between individual fish. This provides a unique form of communication. The distinctive electrical pattern produced by different species of elephant fish, for example, enables them to detect others of the same species, an important consideration when finding a mate in dark and gloomy waters. Within a species, the frequency at which a fish emits is determined by its place in the social pecking order. The higher up in the hierarchy (i.e. the greater the status of the fish) the higher the frequency it uses. This is probably because it costs more energy to produce a higher frequency discharge so that only the 'fittest' fish can maintain their position in the hierarchy. It is the electrical equivalent of the peacock's showy tail.

It is crucially important for a fish to be able to discriminate its own electric signal from those of others in the vicinity. Wave-type electric fish do this by generating signals at a fixed frequency. Each individual emits at its own frequency in much the same way as different radio stations broadcast at different frequencies. However, the number of frequencies is limited so that it sometimes happens that two fish which transmit at the same frequency meet up. This can cause a problem because it becomes unclear which signal arises from which fish, in the same way as it is difficult to distinguish two radio programmes broadcast at the same frequency. Essentially, the fish jam each other's signal, thereby disrupting their electrolocation ability. Should this happen, the fish shift their frequencies relative to one another to maintain their privacy and avoid jamming each other's signals. This separates out the signals emitted by individuals within communication range.

But it is not always sweetness and light and gentlemen's agreements. In a fight, jamming your rival's signal can disorientate it and give you a competitive advantage. Both male and female brown ghost knifefish appear to use such shock tactics when they come into conflict with a rival. Usually if they encounter another fish they switch their frequency to avoid interference, but if they are in competition they deliberately try to jam their opponent's signal and establish dominance. In the social hierarchy of the brown ghost knife fish, the larger and more dominant males emit at a higher frequency and aggressively ramp up their electrical discharge frequency when they encounter a potential rival. This can result in a frequency war, with each fish trying to out-pitch the other's electric signal and so disorientate its competitor.

Amorous male elephant nose fish also use electric signals to lure females. Different species of fish produce pulses that differ in magnitude, duration and frequency, and females are attuned to the signals produced by males of their own species. In some species, complex electrical courtship duets occur, analogous to the courtship songs of birds. Males of some nocturnal gymnotiform fish, for example, serenade their potential mates with long electrical hums and spawning elicits a frenzied electrical extravaganza. It is a costly concert for as much as 20 per cent of the energy consumed by the male fish is used to generate their electrical displays. These megasignals serve to advertise the healthiest males, enabling a female to select the best mate. But this strategy brings a concomitant disadvantage. The electric signals are also picked up by electrosensitive predators, so that the males' numbers are rapidly depleted and few male fish remain by the end of the mating season. To help prevent this decimation, male fish emit high frequency signals throughout the night, when females are more receptive and ready to spawn, but switch to low frequency songs during the day. Sexual strategies appear to be as intricately balanced in male electric fish as in their human counterparts.

7

The Heart of the Matter

Be still, my heart; thou hast known worse than this.

Homer

Early one summer morning, Alex was getting ready for school. Although she was anxious about her exams later that day, she was not unduly stressed and there was nothing that marked the day out as unusual. At least not until she went into the bathroom, reached out to turn on the light – and slid silently unconscious to the floor. Luckily, her mother saw it happen and rushed to the rescue. But this was no simple fainting fit. Alex had a serious heart problem and her increasingly frantic mother was unable to resuscitate her.

By chance, Alex lived close to a fire station and a local fireman picked up the emergency call. He quickly rushed to the rescue and administered cardiopulmonary resuscitation until the ambulance arrived, thus ensuring that her brain and other tissues were supplied with oxygen even though her heart was not working properly and she was no longer breathing. During the journey to the hospital her heart stopped and was restarted more than once. She was unconscious for seventeen hours but eventually recovered.

Subsequent analysis revealed that Alex has an abnormality in the electrical activity of her heart that predisposes her to blackouts and sudden cardiac death. It runs in the family. Her grandmother died in her sleep in her twenties, and her father suffered numerous fainting spells as a child and died young, just a year before Alex's attack. It seems very likely that they carried the same genetic defect that Alex does.

Alex and her relatives are not alone. Other families have experienced similar tragedies, with one or more children or young adults dying in their sleep, when taking exercise, or when stressed. There are even tales of children suddenly collapsing when reprimanded by their teachers, or while running around in the playground. It is not an exaggeration to say that some children with this condition really have died of fright. Happily, our increased understanding of the electrical activity of the heart means that this disease can now be diagnosed from an electrocardiogram or by a simple genetic test, and it can also be successfully treated.

The Beat Goes On

It has been known for centuries that the heart has an intrinsic rhythm and can continue to beat when it is removed from the living animal. One of the first to describe the phenomenon was the great Roman physician Galen, and subsequently many others, including Leonardo da Vinci, reported that the heart moves by itself. William Harvey even showed that when the heart of an eel was cut into ever-smaller parts each individual piece continued to pulsate. This intrinsic activity may have inspired the classical Greek idea that the heart was the seat of the soul. However, the heartbeat has no spiritual origin, but derives instead from electrical events taking place within the cardiac cells themselves.

In essence, your heart is a pump that is controlled by electricity. Blood enters via the upper chambers (the atria), which contract first and force blood into the much larger lower chambers (the ventricles). The ventricles contract in synchrony about half a second later, the right ventricle pumping blood to the lungs and the left ventricle sending it around the body.

Non-return valves lie between the upper and lower chambers of the heart so that the blood only flows in one direction; from the atria to the ventricles. Similarly, non-return valves guard the exits from the ventricles into the great vessels. If these valves leak, as can happen with age, then blood is pumped less efficiently, so that the body receives less oxygen and you feel constantly tired. The chambers on the right and left side of the heart are physically separate which ensures that oxygen-rich blood coming from the lungs is not mixed with oxygen-depleted blood returning from the tissues. However, because heart cells are wired together, they contract in synchrony, so that the heart beats as a single organ.

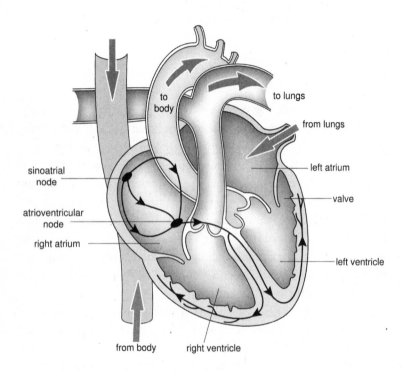

The electrical system of the heart. The pacemaker cells lie within the sinus node in the wall of the right atrium. The black arrowed lines indicate the bundles of fibres that provide the path along which the electrical signals are conducted to the lower chambers (the ventricles). The two sides of the heart are physically separate but contract together. The pulmonary artery carries blood from the right side of the heart to the lungs. Having picked up oxygen there, blood is returned to the left side of the heart, which then pumps it into the aorta and around the body. The period when the heart is contracted is known as 'systole' and the time it is fully relaxed as 'diastole'.

Each heartbeat originates in a group of pacemaker cells (known as the sinus node), which lie in the upper right chamber of the heart. These cells generate electrical impulses that are conducted to the rest of the heart along specialized pathways: first to the atrioventricular node, which lies at the junction between the right atrium and the ventricles, and then to the walls of the ventricles themselves. The time lag associated with electrical transmission ensures that the electrical signals reach the upper chambers before the lower ones, so that first the atria are triggered to contract, and then the ventricles. The timing of this spread of excitation is crucial for the ability of the heart to serve as a pump. If it is disrupted, the heart no longer beats regularly and its capacity to pump blood is compromised.

Although the average heart rate at rest is 70 beats a minute (that's about 100,000 beats a day), it varies widely between individuals. Athletes have much lower resting heart rates, often as little as 40 beats a minute. One of the lowest ever recorded, a mere 28 beats per minute, was that of the cyclist Miguel Indurain, who won the Tour de France five times in a row. By contrast, babies' hearts beat much faster than adults, speeding along at 130 to 150 beats a minute. It turns out that heart rate varies with body size, so that smaller animals (including babies) have higher resting heart rates: the heart of the tiny shrew races away at 600 beats a minute while that of the elephant can only manage a ponderous 25 beats a minute.

The Electrocardiogram

The electrical signals produced by heart cells give rise to tiny fluctuations in the electrical potential at the surface of the body that can be picked up by surface electrodes attached to the skin. This is the basis of the electrocardiogram, commonly abbreviated as ECG (or EKG in the United States).

The electrical activity of the heart was first recorded by Augustus Waller in 1887 both in himself and in his pet dog Jimmie. His demon-

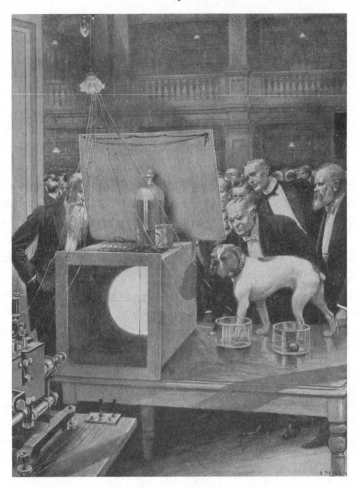

August Waller's pet dog Jimmie was the most popular personage at the annual conversazione of the Royal Society at Burlington House. This scientific party for both scientists and the general public is still held and it traditionally includes many demonstrations. Jimmie stands sedately with his left paws in a conductive salt solution, which is connected to an Einthoven string galvanometer (the large box on the left) that measures his every heartbeat. The string is illuminated by limelight so that its shadow is projected on a sheet, and it vibrates with the bulldog's heartbeat. The experiment is not painful, as many of the audience discovered when they volunteered to take Jimmie's place. August Waller is seen on the far left.

stration of the method at the annual conversazione of the Royal Society of London in 1909, which was open to the public, was reported in the *Illustrated London News*. It triggered a storm of protest in Parliament, with Mr Ellis Griffith, the MP for Anglesey, demanding to know if the Cruelty to Animals Act of 1876 had been contravened. *The Times* stated the Secretary of State, one Mr Gladstone,[1] replied, 'I understand that the dog stood for some time in water, to which sodium chloride had been added, or, in other words, a little common salt. If my hon[ourable] friend has ever paddled in the sea, he will have understood the sensation. (Laughter.) The dog – a finely developed bulldog – was neither tied nor muzzled. He wore a leather collar ornamented with brass studs [this had been referred to by Mr. Griffith in far more emotive terms as "a leather strap with sharp nails [. . .] secured around the dog's neck"]. Had the experiment been painful, the pain no doubt would have been more immediately felt by those nearest the dog. (Laughter.) There was no sign of this.' He might have added that after Jimmie had shown the way, the ladies in the audience queued up to have their heartbeats recorded, by dipping their hands in pots of salt solution and 'their hearts were in every case much steadier than Jimmie's'. As this story also shows, the English concern about animal experimentation has a long history.

Waller's early recordings were of poor quality and unsuitable for clinical purposes, and he is reputed to have said that he did not imagine electrocardiography was likely to find any very extensive use in the hospital and that, at most, it might be 'of rare and occasional use to afford a concrete graphic record of some rare anomaly of cardiac action'. But technical innovations meant that by the 1920s it was routinely used to diagnose heart problems, and it remains an important clinical tool today.

The key was the development of very sensitive instruments, capable of detecting the tiny electric currents produced on the surface of the body when the heart beats. The pioneer in this field was Willem Einthoven, who won the Nobel Prize in 1924 for his invention of the string galvanometer. It consisted of a fine glass fibre, coated with silver to ensure that it could conduct current, which was suspended between two very strong electromagnets. When a current passed though the filament (the 'string' of the galvanometer), the electromagnetic field caused it to move. The greater the current, the more the filament was displaced. This tiny movement was detected by shining a light on the fibre, and the shadow it cast was recorded on a moving photographic plate. All that was needed was to connect the conductive filament to the body. This was done by attaching wires to each end of the filament and immersing the other end of the wires in pots of salt solution. Dipping the hands and feet in the solution completed the electrical connection between the 'string' and the skin. Current from the heart, picked up via the surface of the body, was then able to influence the movement of the filament.

The original string galvanometer was huge. It weighed several tons, took five people to operate and required constant running water to cool the electromagnets. The glass filament, however, had to be very light and thin. It was made by melting quartz glass in a crucible. This was then drawn out into a fine filament by a most unusual means – one more reminiscent of a Boy Scout improvisation than the conventional image of a scientific experiment. The molten glass was attached to an arrow that was then shot across the room, dragging the filament with it and stretching the glass into a very fine 'string'. This was then coated with silver to make it electrically conductive. Safety considerations would undoubtedly ban this experiment today, but fortunately we now have other methods to record tiny currents.

Early photographs show Einthoven sitting with both hands and his left foot (trouser leg carefully rolled up) in separate tubs of conductive salt solution, which were wired up to the monitoring equipment. Today, conducting jelly is used to attach the recording electrodes, one on each arm and one on the left leg. The equipment has also got a lot smaller. Einthoven's original machine occupied two rooms, but nowadays portable twenty-four-hour heart monitors are available that can be worn as the patient goes about their normal life.

The ECG simply reflects the sum of the electrical signals from individual heart cells and it provides a very good, non-invasive indicator of their function. Each ECG complex consists of an initial bump known as the 'P wave', followed by a much larger and sharper bipolar peak known as the 'QRS complex' and then, two to three hundred milliseconds later, by the smaller and slower 'T wave'. The P wave corresponds to the electrical activity of the atrial cells, while the QRS and T waves reflect the beginning and end of the electrical impulse (the action potential) in the ventricular cells. Since these electrical signals drive muscle contraction, the P wave also signifies the contraction of the atria and the interval between the QRS and T waves indicates the duration of ventricular contraction. The delay

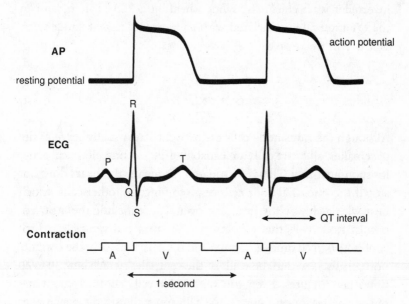

Relationship between the ventricular action potential (AP, upper trace), the electrocardiogram (ECG, middle trace), and the contraction of the heart (lower trace). 'A' indicates the duration of atrial contraction and 'V' that of ventricular contraction. The QT interval reflects the duration of the ventricular action potential.

between the P and Q waves is due to the length of time it takes for the electrical signal to spread from the atria to the ventricles, whereas the interval between the Q and T waves reflects the duration of the ventricular action potential. Why Einthoven should have chosen to name the ECG peaks after the middle letters of the alphabet remains something of a mystery.

The ECG is particularly valuable for detecting irregularities in the electrical activity of the heart and for diagnosing their origin. Changes in the amplitude and timing of the various components can indicate clinical problems. A PR interval that is longer than normal, for example, signals a conduction defect between the upper and lower chambers of the heart known as heart block. An inverted T wave is seen following a heart attack. And an increase in the QT interval is associated with an increased risk of sudden cardiac death.

Sick at Heart

Although the sinus node cells of the right atria usually serve as the pacemaker, all heart cells are capable of generating electrical activity spontaneously. This is fortunate as it means the heart does not stop if the sinoatrial node cells cease to function: other cells, which beat with a slower rhythm, take over. These include the atrioventricular node cells that sit between the atria and ventricle, which contract 40 to 60 times a minute, and the cells that form the conduction pathways within the walls of the ventricles (which beat 30 to 40 times per minute). Even the ventricular cells contract spontaneously. The reason the sinus node cells normally set the pacemaker rhythm is simple: their intrinsic rate is the fastest.

If your heart beats too slowly (a condition known as bradycardia), it will be unable to supply blood quickly enough to your tissues and you will feel tired, weak, dizzy, and short of breath. Walking or climbing stairs becomes a struggle. Tachycardia, when the heat beats too rapidly, is also a problem. At resting heart rates of over 100 beats per minute there is insufficient time for the heart to fill fully between contractions, reducing the amount of blood that can be pumped. Consequently, the tissues will again be short of oxygen and you will be permanently exhausted.

An occasional irregular heartbeat is rather common and many people will have experienced the odd missed beat. In fact, the beat is not really missed – it simply feels as if it is. What actually happens is that a beat arrives early and is not detected because the heart is only half full: there is then an unusually long pause until the next beat, which is more obvious because the heart is then over-filled. Such 'missed beats' are very common, but despite being rather alarming they are of no importance. Although most happen spontaneously, they can also be triggered by stress or drugs such as caffeine.

The most common type of abnormal heartbeat is atrial fibrillation (AF), which affects around 5 per cent of the population over sixty-five. In this condition, the upper chambers of the heart beat erratically and out of synchrony. This happens if the electrical activity of the sinoatrial node cells is disturbed, or if the spread of electrical excitation through the atria is impaired by tissue damage. If the atria beat asynchronously, their ability to force blood into the ventricles is reduced and cardiac output is compromised, causing the patient to feel faint. It also produces a pulse that appears to flutter erratically. Atrial fibrillation can lead to blood clots, which enhance the risk of a stroke, because a clot may lodge in the blood vessels of the brain, cutting off the blood supply to downstream tissues and causing their death (which is why stroke victims may find they cannot speak, or part of their body is paralysed). Normal cardiac rhythm can sometimes be restored by drugs, or by a mild electric shock (a process known as cardioversion), but if atrial fibrillation persists an artificial pacemaker may be needed.

One of the newer treatments for atrial fibrillation is the removal of a small region of atrial tissue, which blocks the circular pattern of electrical activity that underlies the problem. This is usually very effective and recurrence of atrial fibrillation with this treatment occurs far less frequently than with drugs. It can be carried out using a catheter that is inserted into a vein and threaded through the blood vessels to the correct place in the heart. An energy source, such as high frequency radiowaves, is then transmitted via the catheter to selectively destroy the targeted cells.

A more severe condition is heart block, where damage to the conduction pathways means that passage of the electrical signal from the atria to the ventricles is impaired (note it does not mean that the vessels of the heart are blocked). In total heart block, transmission of the atrial signal is completely prevented. Consequently, the ventricles take over, which means the heart rate may fall as low as 30 beats per minute and the patient will have severe difficulty in exercising. Thus an artificial pacemaker is essential.

The most serious arrhythmia of all is ventricular fibrillation (VF), which is fatal if not corrected. In this condition there is electrical chaos, with many regions in the lower chambers of the heart fighting for control of the rhythm. As a result, the ventricles beat so asynchronously that the whole heart appears to quiver continuously, but never contracts properly. It looks, the great sixteenth-century anatomist Vesalius said, like a writhing bag of worms. No significant cardiac output is possible when this happens, so that the heart soon stops through lack of oxygen and the patient dies within minutes. Even before the heart stops, the brain will have been irreversibly damaged by oxygen deprivation. In such a situation, the only hope is to restore normal rhythm immediately. For this it is necessary to stop the heart by administering an electric shock with a defibrillator and hope that it will revert to normal rhythm when it spontaneously restarts – a bit like pressing the reset button on a computer.

A heart attack results from disruption of the blood supply to the heart and is commonly caused by blockage of one of the coronary arteries. As the tissues downstream of the blockage are deprived of oxygen they start to die. This may trigger ventricular fibrillation because the resulting tissue damage prevents the synchronized spread of electrical signals across the heart. Different groups of heart cells then go their own way and start to beat at different times. As in any society, cooperation between the component parts of the heart is vital for it to work effectively.

Restoring the Rhythm

If the heart beats irregularly, an artificial pacemaker is often used to correct its rhythm. Early pacemakers were large and bulky machines, about the size of a washing machine, and they were supplied with mains electricity. Consequently, the patient could not move around easily. They also had another disadvantage: they stopped when the electricity supply failed. In the 1950s, Dr C. Walton Lillehei was doing pioneering open-heart surgery on 'blue babies' at the University of Minnesota. These children were born with a hole between the left and right ventricles, which results in the blood bypassing the lungs, so that oxygen uptake is much reduced. After surgery to repair the hole, some babies suffered from short-term heart block; tissue damage meant that the electrical signals from their sinoatrial node did not reach the ventricles and their hearts failed to beat properly. In such cases, Lillehei used an artificial pacemaker machine until the child's heart healed. This usually took one or two weeks.

Unfortunately, a major power blackout in Minneapolis in October 1957 resulted in the death of one of the 'blue babies'. Infuriated, Lillehei contacted Medtronic, the electronics company that made the machines, and asked for something that ran on batteries. He was in for a surprise, for in less than a month their engineer Earl Bakken returned with an artificial pacemaker that did indeed run on batteries – but it was now shrunk to the size of a sandwich. Transistorized circuits were the key to this miniaturization.

Bakken wrote in his autobiography, *One Man's Full Life*, 'Back at the garage, I dug out a back issue of *Popular Electronics* magazine in which I recalled seeing a circuit for an electronic, transistorized metronome. The circuit transmitted clicks through a loudspeaker; the rate of the clicks could be adjusted to fit the music. I simply modified that circuit and placed it, without the loudspeaker, in a four-inch-square, inch-

and-a-half-thick metal box with terminals and switches on the outside – and that, as they say, was that.' He had intended his prototype as an experimental device for testing on animals and was stunned to discover, when he visited the hospital the next day, that it was already being used on patients. Lillehei calmly informed him that as the device worked he didn't want to waste a minute before using it to help save patients' lives. It was so successful that similar pacemakers were soon introduced throughout the world, and Medtronic became a major supplier.

Just one year later, the first implantable pacemaker was used, in a forty-three-year-old Swedish patient called Arne Larsson. Arne suffered from complete heart block and his death seemed inevitable. His wife, however, had other ideas. She had heard of experiments being carried out on dogs at the Karolinska Hospital in Stockholm and decided that the technology might save her husband. Apparently she was extremely persuasive because she convinced the surgeon Åke Senning and the engineer Rune Elmqvist to help. Rune built the pacemaker in his kitchen. It failed within three hours of implantation, so Arne was given another one the next morning, which lasted just a few weeks. These failures did not put him off, however, and he eventually received twenty-six different pacemakers. The pacemaker enabled him to lead an essentially normal life – and he made good use of it, acting as a patient advisor and advocate for pacemakers throughout the world. He died forty-three years after receiving his first pacemaker, at the age of eighty-six, his bravery and willingness to act as a human guinea pig having doubled his lifespan.

The concept of the artificial pacemaker is very simple. The pacemaker supplies a small electric current which substitutes for the heart's own. This is achieved by inserting a wire into the right ventricle of the heart. It is usually threaded into place through one of the great veins, but in some cases the chest is opened and the wire placed directly on the heart's surface. The lead is then connected to the pacemaker, which applies small electric shocks to drive the heart at the right rate. The pacemaker also contains a battery and, sometimes, electronic circuits that can sense the patient's own heart rhythm and adjust it as needed. Once it is clear the device is work-ing, it is implanted in the chest (usually in front of the shoulder), between the muscle and the subcutaneous fat. The first pacemaker Arne received was the size of a hockey puck, but today they can be as small as a ten-pence piece. They need replacing every five to ten years, depending on how long the battery lasts. As electromagnetic interference can cause pacemakers to malfunction, people with pacemakers must avoid high magnetic fields, cellphones and elec-tronic equipment that generates stray electric fields.

Packer Whackers

Everyone is familiar with the typical emergency-room drama in which the patient is surrounded by a throng of medical staff, des-perately working to save their life. Suddenly, the regular beep of the heart monitor ceases, the normal ECG vanishes to be replaced by a flat line, and someone screams 'Arrest!' Controlled panic ensues. Within seconds, large paddles are slapped onto the patient's chest and with a warning cry of 'Clear!' an electric shock is administered. The patient's chest jerks violently, the heart is restarted and the heart monitor springs into action once again.

But this dramatic scene is far from accurate. There is usually no jerking of the patient in response to the electric shock – this is mere poetic licence. More significantly, in real life an electric shock is not used to restart a patient's heart. Dramatic resuscitations are com-monplace in modern medicine, but they do not occur in patients whose hearts have stopped, but rather in those whose hearts are fibrillating – whose ventricles are beating so asynchronously that the heart is reduced to a twitching lump of flesh that is quite unable to pump blood. And the electric current is not used to start the heart, but to stop it. As previously mentioned, the hope is that when the heart spontaneously restarts, the natural pacemaker cells in the sinus node will take over and the normal rhythm will be restored.

It is possible the popular misconception has arisen from the use of the term 'cardiac arrest'. This does not mean, as might be surmised, that the heart has stopped contracting and is totally still. Rather it refers to the fact that blood flow is arrested. Although individual heart cells continue to contract, they fail to do so in synchrony so that the heart is no longer an effective pump. Within a few minutes the brain dies because of lack of oxygen, and eventually the heart itself ceases to beat for the same reason. Unless the victim suffers a cardiac arrest in hospital, cardiopulmonary resuscitation is required to keep them alive until a defibrillator arrives. Artificial respiration is carried out and the heart is manually compressed by pumping the chest with the heels of the hands, forcing blood out of the heart and around the body. The right speed is vital – too fast and the heart has insufficient time to refill between compressions, too slow and the tissues suffer from lack of oxygen. A rate of 100 compressions per minute is just right. Serendipitously, the Bee Gees' hit song 'Staying Alive' has almost exactly the right rhythm and has been used as a training aid for doctors. Although it also has a near-perfect beat, 'Another One Bites the Dust', by Queen, seems rather less appropriate.

Defibrillators were not commonly carried in Australian ambulances before 1990. That all changed when Kerry Packer, a billionaire well known for his controversial and flamboyant character, had a cardiac arrest while playing polo. By chance, the ambulance that attended the scene was carrying a portable defibrillator. Despite being clinically dead for several minutes, Packer survived. He is alleged to have said of his near-death experience, 'The good news is there is no devil. The bad news is there is no Heaven.' After his recovery, he donated a large sum of money (2.5 million Australian dollars) to equip half the ambulances in the state of New South Wales with portable defibrillators, on the condition that the government paid for the other half. As a consequence, the machines are colloquially known in Australia as 'Packer whackers'. Many Australians owe their lives to his philanthropy.

In recent years, defibrillators have proliferated, and new versions are available that can be used by non-medical operators. In the UK, they are found at railway stations, on airplanes, and in other public places. Although the best-known defibrillators are external devices that are placed on the chest, much smaller implantable devices are also available for people who are at risk of fibrillation. These constantly monitor the heart's rhythm and when necessary deliver an electric shock to reset it back to normal. People with implantable defibrillators can live a normal life secure in the knowledge that they have a built-in 'life-saver'. These apparently deliver quite a shock – it is said to feel a bit like being thumped in the chest.

To Hell and Back

In November 2003, the rock singer Meat Loaf, best known for his performance in the *Rocky Horror Show* and his hit song 'Bat out of Hell', collapsed on stage during a concert at Wembley in front of a large audience. He was rushed to hospital, where he was found to have a rare heart ailment known as Wolff-Parkinson-White syndrome. He later said, 'I remember not being able to sing the lyrics for the song "All Revved Up", walking over to where the girls were and starting to fall.' He thought he'd had a heart attack on stage.

Wolff-Parkinson-White syndrome is a congenital condition that affects between 1 and 3 per cent of the population. It usually only causes problems when the heart rate is very fast, as occurs when someone is stressed or exercising heavily. The sudden unexpected death of very fit athletes due to cardiac arrest, such as that of the ice hockey player Bruce Melanson, is often due to Wolff-Parkinson-White syndrome. Other sufferers have been luckier. LaMarcus Aldridge, an American basketball player with the Portland Trailblazers retired from a game against the Los Angeles Clippers, complaining of dizziness, shortness of breath and an irregular heartbeat. He was subsequently found to have Wolff-Parkinson-White syndrome. Both he and Meat Loaf were successfully treated for the condition.

In the normal heart, electrical signals generated in the atria pass to the ventricles via a specialized conduction pathway known as the atrio-ventricular (A-V) node. People with Wolff-Parkinson-White syndrome have an additional tissue bridge between the atria and the ventricles that provides an alternative pathway for conduction of the electrical signal. The timing of the electrical signal to the ventricles is critical for the heart to beat properly and the A-V node acts as a gatekeeper between the atria and ventricles, modulating the spread of the electrical impulse. If the atria beat too quickly, not all signals will pass through the A-V node which ensures that the ventricles do not beat too fast. The extra conduction pathway found in people with Wolff-Parkinson-White syndrome lacks the special properties of the A-V node and can lead to very fast heart rhythms. It is also possible for the electrical signal to loop around between the atria and ventricles, entering via the A-V node and returning via the additional pathway. This leads to very fast ventricular contraction, which can precipitate fibrillation and sudden death.

Fortunately, Wolff-Parkinson-White syndrome can now be easily cured by a very simple and successful operation in which a catheter is passed into the heart, the offending abnormal tissue bridge identified, and radio frequency pulses used to destroy it.

The Electric Heart

When a heart cell is stimulated it fires off an electrical impulse, or action potential. This spreads rapidly over the surface of the cell and then along a network of fine tubules that penetrate deep into the interior of the muscle fibre. The change in membrane potential to more positive values opens calcium channels within the surface and tubular membranes, so triggering an influx of calcium ions from the extracellular solution. In turn, these serve as intracellular messengers that cause the release of a much larger number of calcium ions from a series of intracellular stores. Interaction of calcium ions with the contractile proteins then causes the muscle cell to shorten. In effect, the electrical impulse is a way of ensuring that calcium increases simultaneously throughout the cell, so that each heart muscle fibre contracts smoothly and synchronously.

As in the case of nerve cells, ion channels are responsible for the electrical impulses of heart cells. However, many more types of channel are involved in shaping the action potential of the heart. It is initiated by the opening of sodium channels. These channels are similar, but not identical to those of nerve cells, which explains why fatal poisons like that of the puffer fish block electrical impulses in the nerves, but not the heart. Errors in the gene coding for the cardiac sodium channel gene (SCN5A) can result in abnormal sodium channels that do not function properly. This gives rise to a rare inherited condition called Brugada syndrome, which disrupts the electrical activity of the heart without warning and can cause sudden death.

Brugada syndrome is most common in the Asian community. It accounts for around 12 per cent of unexplained deaths and – apart from accidents – is the leading cause of death of men under the age of forty in certain regions of the world. Indeed, it is so common in the Philippines that it has a special name – bangungut, which means 'to rise and moan in sleep'. An increased incidence of unexpected death while sleeping is also found in Japan and Thailand (where it is known as Lai Tai, 'death during sleep'). Intriguingly, the disease is far more common in men than women. Perhaps this is why in Thailand it was believed, erroneously, that the disorder could be averted by sleeping in women's clothing. Local superstition has it that young men died because they were snatched away by a widow ghost, who could be tricked into thinking her potential victim was female by their dress. As the ghost was not interested in women, they would escape death.

Understanding the genetic basis of Brugada syndrome came about because of a chance encounter between two scientists who happened to be seated next to one another on the bus ride to the airport following a conference on the heart. When Charles Antzelevitch expressed surprise that no patients had been found with a particular kind of cardiac arrhythmia, his companion informed him that in fact the Brugada brothers had recently described such a rare condition. This fortuitous meeting led to the discovery that Brugada syndrome is caused by loss-of-function mutations in the cardiac sodium channel gene. As many as fifty different mutations are now known to cause the disease. The higher incidence of these mutations in South Asian populations explains the greater prevalence of Brugada syndrome.

Opening of the sodium channel pores is quickly followed by the opening of calcium channels, which enables calcium ions to flood into the cell, where they trigger the release of stored calcium and thereby contraction. The importance of calcium ions for the contraction of the heart was discovered serendipitously by Sydney Ringer in the early 1880s. Ringer was searching for a solution that enabled him to maintain the normal beating of a frog's heart. He did this by adding known amounts of inorganic salts to distilled water, which contains no ions at all. Or so he thought. In fact, because Ringer had a busy life as a medical doctor, the solutions were prepared by his technician, who did not always follow instructions precisely. Ringer's first paper states that only sodium and potassium ions were needed to maintain cardiac contraction. But as he subsequently wrote, 'After the publication' (of his previous paper), 'I discovered, that the saline solution which I had used had not been prepared with distilled water but with pipe water supplied by the New River Water Company. As this water contains minute traces of various inorganic substances, I at once tested the action of saline solution made with distilled water and I found that I did not get the effects described in the paper referred to. It is obvious therefore that the effects I had obtained are due to some of the inorganic constituents of the pipe water.' It turned out that the missing ingredient was calcium – or 'lime' as Ringer called it. One wonders if he praised or castigated his lab technician (probably both).

Calcium channels are not just important for letting in the calcium ions that trigger the release of stored calcium. The fact that they close (inactivate) only slowly at positive membrane potentials helps prolong the cardiac action potential, thereby providing more time for the heart to contract. The action potential of a ventricular cell is about half a second long, almost 500 times longer than that of a nerve cell.

The end of the cardiac action potential is produced by opening of potassium channels, and the resulting efflux of potassium ions returns the voltage gradient across the membrane to its resting value. As a consequence, the calcium channels shut, preventing calcium influx, so that the heart relaxes. Unlike those of nerve cells, many cardiac potassium channels take a long time to open, which helps ensure that the duration of the action potential in the heart is much longer. They also come in several flavours. One of the most important is known as HERG. Its strange name derives from its close relationship to an ion channel in the fruit fly *Drosophila*. This tiny insect is much beloved by geneticists because it has a very fast life cycle, breeds prodigiously, and many genetic mutants have been identified. As flies rarely stay still long enough to be studied, they are usually anaesthetized with ether. In the 1960s, when go-go dancing was all the rage, a mutant fly was found that shook its legs and spun around when exposed to ether, and consequently it was christened ether-á-go-go or EAG for short. Soon after, a related channel was found in the heart and it was named, rather less imaginatively, the ether-á-go-go-related channel or ERG. The human channel thus became HERG.

Frightened to Death

Alex's unexpected collapse one morning from sudden cardiac arrhythmia occurred because she carries a rare mutation in her HERG potassium channel that renders it non-functional. Because these channels are important for ending the cardiac action potential, their loss increases the action potential duration, giving rise to a longer QT interval in the electrocardiogram. For obvious reasons this disease is called long QT (LQT) syndrome. The increase in the QT interval is sometimes very small, a mere 2 to 5 per cent, but it can be sufficient to precipitate a cardiac arrhythmia known as 'torsade de pointes'. The name, which means 'twisting of the points', is taken from a ballet move and describes the distorted form of the ECG. When this happens, the heart no longer pumps blood as effectively and the brain is rapidly deprived of oxygen, which may cause abrupt loss of consciousness. This explains why patients with this condition are prone to sudden blackouts. In some cases, the abnormal electrical activity degenerates into ventricular fibrillation, which can be fatal.

Symptoms of LQT syndrome usually first appear in the pre-teen or teenage years. They are often precipitated by stress, such as exercise, fear or excitement. People have collapsed while running for a bus, diving into a swimming pool, playing in a baseball game, or participating in a TV quiz show. There is usually no warning. Most individuals do not complain of feeling faint or dizzy beforehand; they just abruptly lose consciousness. In about a third of fatal attacks, the person appeared quite fit and healthy before they collapsed, and some people have died while asleep or when aroused abruptly from sleep by the ringing of an alarm clock. Sudden cardiac death was even recognized by Hippocrates, who commented, 'those who suffer from frequent and strong faints without any obvious cause die suddenly'.

Some mutations are particularly severe because they cause deafness as well as cardiac problems: this is because the ion channel concerned is also found in the ear, where it is involved in hearing. An early account of a fatal attack in a patient with this syndrome was given by Meissner in 1856. He vividly described how a young deaf–mute girl who attended the Leipzig Institute collapsed and died while being publically admonished for stealing a small item. Her death made a marked impression on the other children, who saw it as divine punishment for her misdemeanour. When her parents were informed, they were not surprised. It turned out that there had been similar tragic incidents in the family beforehand – one child had fallen down dead after sudden shock, and another after a terrible tantrum.

The death of a young child is a heartbreaking event but it is especially devastating when a seemingly healthy baby unexpectedly dies while asleep. In such circumstances, foul play may be suspected, adding to the agony, and it is not unknown for cot death to result in prosecution of the parents and a murder conviction. Even when this is not the case, not knowing the cause of your child's death can be a lifelong burden. Recently, it has been found that some cot death victims carried ion channel mutations that predisposed them to LQT syndrome, suggesting that they may have died of sudden cardiac death. Just how many cases of cot death are caused from heart arrhythmias precipitated by defective ion channels is still unclear. Nevertheless, post-mortem screening for ion channel mutations would seem a good idea, not only to help identify the cause of death, but also because of the possibility that other family members may be asymptomatic carriers of any mutation identified and thus potentially at risk.

Fortunately, LQT syndrome can now be treated, enabling patients to lead a relatively normal life. Drugs known as beta-blockers prevent the effects of stress on the heart and are usually highly effective. Many people are also given an implantable defibrillator, which can detect an abnormal heart rhythm and apply a shock to reset it back to normal.

The Tale of Terfenidine

Mutations in many different genes are known to cause LQT syndrome, including at least six different kinds of ion channel (most are potassium channels). But LQT syndrome is not always genetic in origin. It can also be induced by drugs that block cardiac ion channels. The drug terfenidine is a very effective anti-allergy agent that used to be sold over the counter in the UK. In 1985 it was approved for use in the USA, where it was marketed as Seldane. It quickly became widely accepted and by 1991 it was the ninth most prescribed medication in the United States. But by that time several cases of cardiac problems in patients taking the prescribed dose of terfenidine had been reported, including prolongation of the QT interval and sudden death. In most cases, affected patients were also taking certain antibiotics, had liver dysfunction, or had pre-existing cardiovascular disease. Consequently, the pharmaceutical company who manufactured the drug issued 1.6 million letters to physicians and pharmacists recommending it should not be used in patients with these conditions. It was later removed from the market.

Terfenidine has this effect because it blocks HERG potassium channels. In most people terfenidine does not produce problems as it is rapidly broken down in the liver to a metabolite that does not block HERG, but remains an effective anti-allergy agent. As the drug is taken orally, it passes through the liver first, so little terfenidine ever reaches the heart. However, individuals with liver disease, who may be deficient in the enzymes that break down the drug, or people taking drugs or substances (such as grapefruit juice) that inhibit these enzymes, are at risk of developing cardiac arrhythmias.

The tale of terfenidine does not end here. It was quickly recognized that many other drugs were also able to block HERG and thereby predispose people towards heart problems. Thus in 2001 Japan, the USA and the European Community ruled that all novel drugs must be screened for their effects on HERG. The most recent guidelines dictate that studies must be performed not only on isolated cells and tissues but also on people (thousands of ECGs are required). This ruling led to a plethora of small biotech companies that carry out HERG screening and to a sharp rise in the cost of drug development because many drugs fail at this stage. Some drug companies who had drugs in the later stages of the development pipeline that turned out to interact with HERG lost very considerable amounts of money.

My Heart Goes Pit-a-pat

Her: Oh doctor, I'm in trouble.
Him: Well, goodness gracious me!

Her: For every time a certain man
Is standing next to me.
Him: Mmm?
Her: A flush comes to my face
And my heart begins to race,
It goes boom boody-boom boody-boom boody-boom
Boody-boom boody-boom boody-boom-boom-boom.

So begins the hilarious duet famously sung by Sophia Loren and Peter Sellers. It's a familiar feeling: all of us have experienced the speeding of the heart when we are excited or afraid, and the thumping beat that makes us feel as if our heart is about to burst.

This is caused by the 'fight or flight' hormone adrenaline, which primes the body to cope with an adverse situation by increasing both the rate and the force of contraction. It does so by opening additional calcium channels in heart cell membranes. This speeds up the rate at which the sinus node cells fire, so that the heart rate is increased, and it also boosts the amount of calcium that is released from the intracellular stores and thereby enhances the strength of contraction. Adrenaline is made by the adrenal glands that lie just above the kidney, and is secreted into the bloodstream in response to stress or exercise; a related substance with a similar action, noradrenaline, is released from nerves that innervate the heart.

Although an increased heart rate during exercise is essential in order to ensure that the limb muscles are adequately supplied with fuel and oxygen, too fast a rate is deleterious. This is because the heart muscles themselves cannot be supplied with oxygen fast enough. The consequence is angina – a severe incapacitating chest pain that can extend down the left arm. Angina is more easily precipitated in people whose coronary blood vessels are narrowed as a result of atherosclerotic plaques (fatty deposits in the vessel walls). Consequently, an exercise test, which increases the heart rate and thus its oxygen demand, is often used to test the health of the coronary vessels. Angina is not only brought on by exertion: it can also be triggered by anger, excitement or emotional stress. I vividly remember the time I was sailing a small yacht up the Ijmuiden canal to Amsterdam and debris became entangled around our propeller, rendering the engine useless. This canal is a major shipping route and it was extremely busy. Huge, heavily loaded commercial barges, with very limited ability to manoeuvre, were bearing down on us. As I struggled to put up the sails and the other crew member dived overboard with a knife to free the propeller, the skipper had an angina attack. He retired below deck to crush a glass capsule of amyl nitrate (nitroglycerin) under his nose and inhale the vapour. This eased his pain by dilating the coronary vessels and increasing blood flow to his heart.

Nitroglycerin acts by releasing a natural gas called nitric oxide, which stimulates the production of a chemical called cyclic GMP that causes blood vessels to relax. Viagra (sildenafil citrate) works in a similar way: by elevating cyclic GMP levels in the blood vessels of the penis it causes them to dilate, resulting in an erection. However, if both drugs are taken together their effects can summate, causing blood vessels throughout the body to relax so much it leads to a severe drop in blood pressure. Consequently, men taking nitroglycerin for their angina should avoid Viagra. It is a fascinating fact that Viagra was discovered fortuitously by scientists seeking drugs to treat angina. It wasn't very effective in clinical trials and would have been abandoned, but for the fact that a few men taking part in the trials were reluctant to stop taking it because of an unusual (and unexpected) side-effect.

Beta-blockers are often taken to decelerate a racing heart. They work by blocking the action of adrenaline, preventing it binding to beta-adrenoreceptors in the heart membrane and so speeding up the heart. However, they can have an unfortunate side-effect, as some men quickly find out when the drug renders them impotent.[2] The incidence is relatively low, although, interestingly, some studies suggest that it is higher in men who are aware of this side-effect of beta-blockers, suggesting that at least part of the problem is caused by anxiety. One case, perhaps, where too much knowledge may indeed be a dangerous thing.

Be Still, my Heart

Chemicals released by the nerves innervating the heart can also slow its rate, and sometimes even stop it completely. In 1994, I was visiting Houston in Texas for a scientific conference. It had been a long and tiring flight and it was terribly hot, but I was determined to go to the welcome party. I had had one (well, maybe two) glasses of wine when suddenly I felt wobbly, dizzy and my head seemed to explode. The next thing I remember was staring down a black tunnel at an enormous polished mound, which I slowly became aware was the tip of a man's shoe. And then there were many of them, filling my vision. I was on the floor – cold, sweaty and with a mouse's-eye view of the feet of my colleagues. I had fainted for the first time in my life. The reason was simple: a sudden increase in the activity of the inhibitory nerves supplying my heart had temporarily stopped it. Consequently, my brain ceased to receive any oxygen and I blacked out. Once on the floor, however, with the blood supply restored, I revived.

The chemical transmitter acetylcholine is responsible for slowing the heart rate. It is released from the terminal branches of the vagal nerve, which runs from the brain to the heart (among other organs). Acetylcholine binds to muscarinic receptors on the sinus node cells. These receptors derive their name from the fact that they are also activated by muscarine, a compound found in certain mushrooms, including the familiar red-and-white-spotted flycap *Amanita muscaria*. Binding of acetylcholine to muscarinic receptors (which are different from the acetylcholine receptors found in skeletal muscle) triggers a chain of reactions that ultimately leads to opening of potassium channels. This allows potassium ions to flow out of the cell, making the interior of the cell more negative. As in the case of the nerve cells, this switches off the sodium and calcium channels, so decreasing electrical activity and slowing the heart rate.

The heart is under a small but continual amount of inhibition by the vagal nerve, which explains why the resting heart rate is actually slower than the spontaneous firing rate of the pacemaker cells in the sinus node. People who have had a heart transplant lack any nervous input as the vagal nerve is severed during the operation and so their resting heart rate is higher than normal.

Atropine antagonizes the action of acetylcholine at muscarinic receptors and is used clinically to reduce the effect of the transmitter in patients with a very slow heart rate, or whose heart has actually stopped. This helps speed up the heart. In large amounts, however, atropine is a deadly poison. It is named after Atropos, the most terrible and feared of the three Fates in Greek mythology, who cut the thread of life, and whose hand could never be stayed.

Atropine also inhibits muscarinic acetylcholine receptors in other tissues. One of its most celebrated effects is to dilate the pupil of the eye. Brilliant eyes, due to dilated pupils, are perceived as more sexually attractive, perhaps because orgasm also induces widening of the pupil. Atropine was widely used as a cosmetic by ladies of the Elizabethan court who obtained it by crushing the shiny black berries of the deadly nightshade plant, which explains its Latin name – *Atropa belladonna* (meaning 'beautiful lady'). Every part of the plant is poisonous to humans, although birds can eat the seeds with impunity. Atropine and its derivatives are used clinically today to dilate the pupil in eye examinations and enable the ophthalmologist to examine the back of the eye more easily. You may even have experienced its effects yourself – this is the drug that makes your eyes super-sensitive to light (because the iris muscle can no longer contract in bright light), so that you tend to screw up your eyes in the sunshine and should not drive.

A Racing Heart

You have only to run for the bus to recognize that exercise has a dramatic effect on the heart rate. The maximum human heart rate is around 200 beats a minute, about threefold greater than at rest. It can be very much higher in other creatures – that of the hummingbird during flight is a staggering 1,200 beats a minute. This increase in rate is triggered by the release of noradrenaline from the sympathetic nerves innervating the heart and by a rise in the circulating levels of adrenaline. Although people with heart transplants increase their heart rate in response to exercise, they do so more slowly as they respond only to adrenaline in the blood, which takes longer to get there. The brake on the heart rate produced by acetylcholine released from the vagal nerve is also removed during exercise and reinstated when exercise ceases: this does not happen in transplant patients, which explains why their heart rate takes longer to return to normal after exercise.

The maximum heart rate is age-dependent (decreasing with age) but similar in all people, independent of their fitness. What varies is the maximum amount of blood they can pump. Athletes have lower resting heart rates because regular exercise produces an increase in the size of the heart and thus enhances the amount of blood that it can pump with each beat. Consequently, the heart needs to beat less frequently to pump the same amount of blood around the body. Although their maximal heart rate is similar, athletes are able to pump far more blood when exercising than couch potatoes because their hearts are larger, giving them a competitive advantage.

The Silent Killer

Potassium chloride is a very effective way of stopping the heart. It is fast, silent, leaves little evidence behind, and is said to be painless (although who is telling?). It is thus a favourite method of murder in detective stories, such as in the Dick Francis novel, *Comeback*, in which horses, and humans, are killed by infusing them with a potassium chloride solution. In *Comeback*, the chemical is obtained from a specialist company, but it is actually very easy for anyone to get it, because it is widely sold as a low-sodium salt substitute. Nor is murder with potassium chloride confined to fiction: a number of hospital nurses have been charged with, and even convicted of, unlawful killing of patients in their care by injections of potassium chloride.

Intravenous injections of potassium chloride, following an anaesthetic to put the victim to sleep, have also been used legally in state
executions of criminals. Dr Jack Kevorkian famously used them in
his thanatron machine,[3] a euthanasia device he used to help termi-
nally ill patients die (he was jailed for second-degree murder in 1998)
and rather improbably, potassium chloride has also been promoted
as a self-administered suicide aid by the German ex-politician Roger
Kusch.

Why does potassium chloride stop the heart? At high concentrations, it depolarizes the heart cells so much that the sodium and
calcium channels are switched off (inactivated). Because these pores
are shut, no action potentials are generated, so that the heart simply
stops. If potassium is infused slowly, however, it is likely that the
heart will first speed up, and then enter ventricular fibrillation
before stopping.

Interestingly, potassium levels in the blood rise during exercise,
due to the release of potassium ions from working muscle. In heavy
exercise, the level attained would be sufficient to stop the heart at
rest. Yet few people's heart stops when they run. It is not fully
understood why this is the case, but one possibility is that it is due
to a protective effect of the hormone adrenaline, which also rises in
exercise. If the blood potassium concentration does not come down
fast enough after stopping exercise then the person may suffer post-
exercise collapse. This could account for the fact that it is more
common to suffer a heart attack shortly after finishing a squash
game than when you are actually on the court.

The Virtual Heart

We now know most of the different kinds of ion channel that contribute to the electrical activity of the heart. There are very many of
them. Different types of heart cell may have a different complement
of ion channels, and the density and activity of the same kind of
channel can vary depending on where the cell is located in the heart.

Thus is it very difficult to predict what will happen to the electrical activity of a single cell when a specific ion channel is modified, let alone what happens to the electrical activity of the whole heart. For this, a computer model is essential.

A key aim of current cardiac research is to develop a real-time computer model of the heart. The principal exponent of this approach has been the Oxford professor Denis Noble. His 'virtual heart' model is good enough to model the normal heartbeat, the effects of a heart attack, of genetic mutations that cause human disease, and the actions of drugs that block HERG channels. It is even sometimes used by drug companies to help explain the mechanism of action of novel drugs.

Some years ago, when his model was in an earlier version, the pharmaceutical company Roche asked Noble to appear in person at a hearing of the Federal Drug Administration in Philadelphia. To his surprise, he found the back of the hall was filled with traders clutching mobile phones and listening to the proceedings. The price of Roche stocks on Wall Street rose and fell as new evidence was presented and forwarded to the New York stock exchange. After the professor's presentation, one member of the FDA remarked, 'I want my hands on that program.' 'No problem', came the laconic reply, 'but you will need to buy a supercomputer costing £5 million [£10 million in today's prices] to run it.'

Computing power has increased so rapidly that today the same simulations can be run on an ordinary desktop computer. But to simulate the activity of the heart in real time, as the drug companies desire, is still beyond the capability of most modern supercomputers (at least for now).

8

Life and Death

Life and death are balanced on the edge of a razor.

Homer, *The Iliad*

In 1970, about 15 per cent of the maize (corn) crop in the United States succumbed to an epidemic of Southern Corn Leaf Blight, caused by the fungus *Bipolaris maydis*. An estimated one billion bushels of maize were destroyed at a cost of over a billion dollars, and many small farmers went out of business. The disease was first reported in the USA in 1969, but cases were isolated and it was not considered of major importance. That all changed in 1970. The warm humid weather that year provided ideal conditions for the rapid spread of the fungus. The epidemic started in Florida and by June it had reached Alabama, southern Louisiana, much of the Mississippi basin and parts of Texas. By September the disease had spread throughout the Corn Belt and as far as Wisconsin in the north and Kansas in the west.

It caused extensive devastation of the corn crop. The first signs of infection were a reddish discoloration of the leaves, rapidly followed by yellowing of the whole plant. In the worst cases, the ears of corn rotted and fell to the ground, where they crumbled into pieces. Some fields were so heavily infested that when they were harvested clouds of spores boiled blackly above the machines.

The destructive effects of Southern Corn Leaf Blight resulted from the unhappy combination of a toxin produced by the fungus and an ion channel that is only found in certain self-sterile strains of maize. The disease reached epidemic proportions in 1970 because most of the maize planted that year was of the self-sterile variety. The reason for this genetic uniformity has its origins in the 1800s. Like many plants, maize is a hermaphrodite and has both male and female parts. The male parts are the tassels, which stick up from the top of the plant and release pollen grains into the air. The female parts are found in the ear of the corn; they develop into the maize kernels following fertilization. Wild maize is self-fertile and, because of the proximity of the male and female parts, most plants are self-fertilized. However, the best maize is a hybrid, obtained when the female plants are fertilized by pollen from a different strain. This was discovered toward the end of the nineteenth century, when plant breeders found that hybrid plants tended to be taller and more vigorous than either of their parents and, more importantly, had larger ears and kernels. Gradually, the use of hybrid maize spread. The farmers were impressed by the superior properties of the hybrid seed and the seed merchants who supplied them encouraged the use of hybrid strains, as it meant that farmers had to buy new seed each year.

To obtain hybrid plants it is necessary to prevent self-fertilization. Historically, this was done by removing the tassels by hand, a time-consuming and laborious operation, since it had to be done for many thousands of plants each year. Crop breeders were therefore delighted when they came across varieties of maize that produced no pollen. They realized immediately that these plants, known as cytoplasmic male sterile (CMS) strains, would be ideally suited for cross-breeding and they were soon widely adopted by seed companies. All that was necessary was to plant the CMS strain adjacent to a pollen-producing variety of maize and the wind would do the rest: the CMS plants would produce only hybrid seed. However, there was a hidden cost. Unlike normal maize plants, and unbeknownst to the plant breeders, the male-sterile CMS varieties were susceptible to Southern Corn Leaf Blight because they carried a specific type of ion channel in all of their cells.

As this story illustrates, ion channels are not limited to electrically excitable cells like nerve and muscle. They are found in every cell of our body and in every organism on Earth, from the humblest bacteria to the giant redwoods of California, and they regulate everything we do.

Turbo-charged Sperm

Ion channels play a crucial role in our lives even before conception, for they influence the outcome of the great sperm race. An arduous event with only one winner, it is the first and most important race we will ever enter and one that each of us (or rather some part of us) has won.

Sperm must swim from the moment of ejaculation, fighting their way towards the egg by lashing their tails. As they travel from the vagina into the upper regions of the woman's reproductive tract they encounter a more alkaline environment and the hormone progesterone. This triggers a switch in the beat of the sperm's tail from rapid wriggles to slower, larger and more forceful asymmetric whips that spur on the sperm. It's a kind of last-minute turbo charge that kicks in just when the sperm needs more power, and it is essential – without it, the sperm would lack the thrust to penetrate the membranes surrounding the egg. This change in the beat of the sperm's tail is produced by the opening of a specialized ion channel known as Catsper.

Catsper is the favourite channel of David Clapham, a Harvard scientist with a razor-sharp brain, a wicked grin and a salacious sense of humour. Searching through the database of the Human Genome Project for undiscovered treasures, his post-doctoral fellow Dejian Ren came across a novel ion channel that was found only in the testis. That fact immediately caught Clapham's attention and soon sperm, in all their many manifestations, became a focus of the lab. 'They have', Clapham says, 'everything neurones have and more: they have ion channels, they get excited, they sense chemicals in their environment, they move – and they get more vigorous around an egg, just like men around women.'

The Catsper channel is one the most complex in the human genome. The channel pore is composed of four different proteins and it associates with several different kinds of accessory protein. If any one of them is absent, the channel no longer functions and the sperm fails to switch to the stronger tail lashes, resulting in infertility. As Catsper is only found in sperm, drugs that block the channel might make a perfect contraceptive. Unlike the more familiar contraceptive pill, they would not interfere with a woman's hormonal system and they would not need to be taken orally. But such a drug would not be the long-sought male contraceptive. It would again be the woman who would have to take it, not just so she could be confident of its use, but because it is only in the her reproductive tract that the change in the sperms' swimming occurs.

Not all sperm have Catsper channels. They are not found in the impressively giant sperm of the tiny fruitfly *Drosophila bifurca*, which crawl rather than swim up the female tract. These leviathans have the longest tails on Earth. They are almost six centimetres (over two inches) long, which is more than 600 times longer than their human counterparts and 20 times as long as the fly itself. Why such giant tails have evolved remains a mystery, but one idea is that the curled-up tail forms a plug that completely fills up the female tract and prevents other sperm from entering. Competition between sperm to pass on their DNA is intense, even when they come from the same male.

Plants have a different problem, as their sperm are immotile and contained within pollen grains to protect them from desiccation. Yet they too use ion channels to facilitate fertilization. When a pollen grain lands on a plant's female reproductive organ (the stigma) it sends out a long pollen tube that grows down towards the egg, carrying the sperm within it. The tube bursts on arrival, releasing the sperm. It turns out that the rupture of the pollen tube is caused by a chemical secreted by the cells surrounding the egg that opens an ion channel in the pollen tube membrane. As a consequence, potassium ions flood in, dragging water with them and causing the pollen tube to swell and burst. Liberated from the confines of the pollen tube, the sperm can now fertilize the egg.

Raising the Barriers

It is vital that only a single sperm fertilizes an egg because if multiple sperm do so the resulting cell fails to develop normally. Thus the egg has developed defences to ensure that only the first sperm to arrive is welcomed and that all subsequent hopefuls are excluded. How this block to polyspermy is produced was first studied in sea urchin eggs, which are easier to work with, as they are very large and can even be seen with the naked eye. Back in 1976, while still a young student, Rindy Jaffe discovered that as soon as the first sperm penetrates a sea urchin egg, the potential across the egg membrane rapidly flicks from being negatively charged on the inside to being positive. This voltage difference prevents further sperm from entering.

The surprise came when scientists looked at mammalian eggs and discovered that the mechanism was different. Here, the block to polyspermy is not an electrical but a physical one – a mechanical barrier that the sperm cannot penetrate, which develops only slowly after fertilization. The difference in strategy reflects the very different environments in which fertilization takes place. In the ocean, many millions of sperm arrive almost simultaneously at the egg so an electrical block to polyspermy is ideal as it is very fast. In mammals, the long and difficult journey up the female tract ensures that only a few sperm make it to the egg and that they rarely do so simultaneously. Hence a slower block is adequate.

Drawing Life from Death

Aya Soliman had a most unusual start in life, being born by Caesarean section two days after her mother Jayne was declared brain dead. Jayne, a champion ice skater, had a fatal brain haemorrhage when she was twenty-five weeks pregnant. She was flown by air ambulance to hospital in Oxford, but died shortly after arrival.

Although Jayne's brain was dead, doctors decided to keep her body alive to provide vital time for her daughter's lungs to mature.

Within the womb, the fetus floats in a cushioning sac of water. Its developing lungs are filled with fluid and it does not breathe air, but obtains all the oxygen it needs via the umbilical cord that links it to the placenta. At birth, the water within the lungs must be rapidly removed as the newborn child switches over to breathing air. This is achieved with the help of specialized epithelial sodium channels (ENaC channels) that are present in the cells that line the lung. At birth the ENaC channels open, allowing sodium ions in the lung fluid to flow down their concentration gradient into the lung cells. Because sodium ions drag water with them, the lungs quickly dry out and so long as ENaC channels are present and functional, the lungs are rapidly cleared of fluid. Without ENaC, however, babies are at risk of drowning in their own fluid at birth and may suffer from 'wet' lungs.

During normal development, a rise in steroid hormones switches on ENaC production a few weeks prior to birth, ensuring the lungs are fully mature when the baby is delivered. At twenty-five weeks of pregnancy, however, lung development is incomplete and the number of ENaC channels in the cells lining the lung is still very small. A chemical called surfactant that reduces the surface tension of the tiny air sacs in the lungs and so prevents their collapse is also low. Thus if a baby must be delivered early, and conditions permit, steroids are administered to the mother before birth. These cross over the placenta and help her premature baby's lungs mature. As a mother's womb is the optimal incubator for a baby, Jayne's body was kept alive on a life-support machine while steroids were given to provide her daughter with the best possible chance of life.

There is a further twist to this story. It turns out that at birth ENaC channels are stimulated to open more completely by the stress hormone adrenaline, which rises dramatically in the mother's blood during the trauma of labour. This may explain why babies born by Caesarean section, where this stimulus is lacking, may have more difficulty clearing their lungs than those born naturally, and why they experience a higher incidence of respiratory complications in the postnatal period.

Piling on the Pressure

ENaC's tasks do not end at birth. It plays a vital role in regulating the amount of sodium in your blood and this, in turn, determines your blood pressure. If ENaC channels malfunction, your blood pressure can skyrocket, putting you at risk of a stroke.

Your kidneys are sophisticated machines that clean the blood, continuously filtering out toxins and waste products and flushing away excess water. Waste processing takes place in about a million individual units known as nephrons, where tufts of fine blood vessels, known as capillaries, are entwined with tiny tubules that act as urine-collecting devices. Amazingly, the whole of your blood passes through the kidney twice every hour. The red blood cells and plasma proteins are retained in the capillary, but the salts and water are forced out into the kidney tubule. Almost all of the sodium and much of the water that is filtered are subsequently reabsorbed as the fluid passes down the kidney tubules. What remains is stored in the bladder and excreted as urine.

ENaC channels in the membranes of the kidney tubule cells are responsible for reabsorption of sodium. As in the lung, sodium uptake is accompanied by water, which leads to an increase in blood volume and, because the circulation is a closed system, raises the blood pressure. A diet high in salt (sodium chloride) is bad for you because more sodium is taken up, which drags more water with it, increasing your blood volume and therefore your blood pressure. Conversely, if blood sodium levels are low, insufficient water is retained by the body, leading to a fall in blood pressure. This is why it is important to ensure that you eat enough salt in a hot climate, where a lot of salt is lost through sweating.

Mutations in any of the three genes that make up the ENaC

channel affect blood pressure. Those that lead to increased ENaC activity cause a hereditary form of hypertension known as Liddle's disease, whereas those that reduce ENaC activity result in low blood pressure. The latter are particularly dangerous as they lead to a life-threatening salt-losing syndrome in newborns and infants. Because sodium uptake is reduced, less water is reabsorbed, so that the child quickly becomes dehydrated and the blood concentration of other ions (especially potassium) becomes unbalanced. The disease is fatal unless it is quickly recognized and treated.

Fortunately, mutations in ENaC are rare. However, it is thought that one reason for the greater incidence of high blood pressure and its attendant complications in black people than in Caucasians is because they have relatively common variants in their ENaC channel genes that predispose them to increased sodium uptake. Why this is the case is uncertain, but one suggestion is that people living near the Sahara evolved very efficient mechanisms for absorbing salt as it was in such short supply. While this is advantageous when salt is only rarely obtainable, it becomes a handicap in our present world where much processed food is very high in salt.

A Salty Tale

In mediaeval times, kissing a child's forehead was held to foretell its fate, for those that tasted salty were considered to be bewitched and at risk of dying young. The association between a salty skin and an early death is not mere myth, however, but an early description of the disease we now call cystic fibrosis. People with this disease have very salty sweat and fail to secrete certain digestive enzymes. Most serious of all, their lungs get clogged up by thick, sticky, mucous secretions that make it hard to breathe and lead to chronic infection, inflammation and a slow progressive destruction of the lungs. There still is no cure. It is a life-threatening condition and even with today's advanced technologies over half of those born with the disease are likely to die before they reach the age of forty.

Cystic fibrosis was first recognized as a distinct disease in 1938, when Dorothy Andersen published the first comprehensive description of the disorder. Several years later, during a heatwave in New York City, the paediatrician Paul di Sant'Agnese noticed that many children admitted to hospital with heat prostration also suffered from cystic fibrosis. An insightful physician, he recognized that their collapse was probably precipitated by excessive salt loss and he analysed their sweat. It contained an abnormally high level of sodium chloride, a finding that forms the basis of the 'sweat' test that is still used today in the diagnosis of the disease.

Mutations that ablate the function of an unusual ion channel lie at the heart of the problem. Its full name is the cystic fibrosis transmembrane conductance regulator, but as this is such a mouthful it is always abbreviated to CFTR. This channel resides in the cells that line the lungs and the ducts of organs such as the sweat glands, pancreas and testes, and it shepherds chloride ions across the cell membrane. Secretion of chloride ions is vital for formation of the fluid that carries the digestive enzymes into the gut, for the seminal fluids, and for sweat. It is also essential for fluid secretion in the lungs, where a thin film of fluid is used to entrap bacteria and move them from the base of the lungs up the airways to the mouth, where they are swallowed and safely destroyed. Without this escalator, the airways clog up with a thick, sticky mucous in which bacteria breed. Such infections eventually damage the lungs.

Current treatments for cystic fibrosis involve simply managing the symptoms: fighting lung infections with antibiotics, preventing the build-up of mucous in the lungs by physiotherapy, and replacing the missing digestive enzymes. But new research aims to correct the defective channel itself. About 4 per cent of patients have a mutation in CFTR (known as G551D) that reduces the time the channel spends open. Recently, a drug known as ivacaftor has been shown to coax such sleepy channels into functioning normally, and preliminary studies suggest it may be of clinical benefit. While there is still a long way to go, it is a promising new approach for treating people with the G551D mutation. Most patients, however, have a different variant of CFTR, known as F508del, that prevents the channel from ever reaching the surface membrane of the cell. In this case, drugs that correct the defective targeting of the channel are needed.

Cystic fibrosis is extremely rare in Orientals and black Africans and is highest in individuals of northern European extraction, where it is one of the most common inherited single gene diseases. Around 9,000 people in the UK have the disease and one in twenty-five of the population – over two million people – carry one copy of the faulty gene: although they are asymptomatic themselves, when two of them have a child there is a 25 per cent chance the child will have cystic fibrosis. This high frequency suggests there may be a selective advantage in having a single copy of the gene. One possibility is that such 'carriers' may be more resistant to the effects of diarrhoeal diseases, such as cholera. *Vibrio cholerae*, the bacterium responsible for cholera, produces a toxin that leads to opening of CFTR channels in gut cells, so that chloride rushes out of the cell, dragging water with it. This causes massive fluid loss from the gut, which results in severe diarrhoea and rapid death from dehydration. Individuals with a lower complement of CFTR channels may secrete less chloride and thus potentially be less susceptible to dehydration.

The cholera bacterium is transmitted in faeces, and any natural disaster that leads to a breakdown of sanitation, such as an earthquake or floods, brings with it the risk of a cholera outbreak. The 2010 earthquake in Haiti was no exception and was quickly followed by an epidemic of the disease. Although cholera is no longer a disease of northern Europe, being mainly confined to third-world countries, this was not always the case. One of the most notable successes of public hygiene was the removal of the handle of a water pump in Broad Street, London, by Dr John Snow in the summer of 1853.

During a severe fourteen-week-long cholera outbreak, Snow noticed that there were about ten times as many deaths in the district of Southwark than in Lambeth. He was of the opinion that cholera was spread in the water, whereas others contended it was the 'foul miasma' seeping from the sewers. Diligent study led Snow to discover that in one area of London the pipes from two different water companies were intermingled so that people were exposed to the same air and environment, but not necessarily the same water. By removing the handle of the pump that supplied infected water, he contained the outbreak of cholera and confirmed his hypothesis that the disease was spread in the water supply. The outbreak was eventually traced to Frances Lewis, a five-month-old baby who died of an attack of violent diarrhoea. Her mother poured the water used to rinse her daughter's infected clothes into the gutter outside her house, which leaked into the Broad Street well and contaminated the water supply. It was a fatal mistake.

The Cell's Plumbing System

Ultimately, both ENaC and CFTR produce disease by affecting transcellular water fluxes. For many years scientists puzzled about how water could cross cell membranes. As they are made of lipids (fats), they should be largely impervious to water, so how was it possible for water to penetrate the lipid barrier in such large amounts that it produces tears, saliva, sweat and urine? The answer is that most cells have specialized water channels known as aquaporins that conduct water across the membrane into and out of the cell. They were discovered serendipitously by Peter Agre. He called his finding, which eventually led to the award of a Nobel Prize, 'sheer blind luck'. Having a suspicion that the protein he had discovered might be the long-sought cell water channel, he tested its ability to transport water using frog eggs, which normally live very happily in freshwater. To his excitement, frog eggs engineered to express water channels in their membranes swelled up and burst when they were placed in freshwater.

Agre's experiment was a perfect demonstration of the power of osmosis – the tendency of water to flow from a region of low salt concentration to one of a higher concentration. Because freshwater has far fewer salts than those inside the cell, water will always attempt to penetrate frogs' eggs but it is normally prevented from doing so by the lipid membrane. Increase the water permeability of that membrane, however (for example, by adding lots of water channels as Agre did) and water will rush in, causing the egg to swell and eventually burst.

It turns out that there are many different kinds of aquaporin channels and they are present in many types of cells, including brain cells and red blood cells, and even the cells of plants and micro-organisms. One of the most important (known as aquaporin 2) sits in the collecting ducts of the kidney tubules and is responsible for reabsorbing the final thirty-five litres of water that the kidney filters every day, and thus for our ability to make a concentrated urine.[1] Approximately three billion water molecules a second pass through a single aquaporin channel. It is highly selective as, due to the unique architecture of the pore, only water – and not ions – can pass through. Water channels are also unusual in that they do not open and close like ion channels, but are permanently locked open: instead the amount of water taken up is regulated by shuttling the channels in and out of the cell membrane. When the body needs to conserve water, extra water channels are inserted. Conversely, if you drink too much fluid, water channels are removed, so that less of the water filtered by the kidney is reabsorbed and it simply flows away as urine. This endless cycling of water channels into and out of the cell membrane is under hormonal control and occurs continuously. It is happening in your own kidneys, right now.

Interestingly, the process can be disrupted by alcohol. A few pints of beer prevent the release of the anti-diuretic hormone that causes water channels to be inserted into the kidney tubules, which is why you produce copious amounts of dilute urine. The result is that the morning after a binge you wake up in a partially dehydrated state, which contributes to the headache. As all the alcohol has now been metabolized (one hopes), hormone levels will be higher, water channels will be mobilized into the tubule membranes, and the increased water uptake will result in a concentrated urine. You can even observe the phenomenon for yourself for the concentrated urine you produce the morning after an evening out is a far darker colour than the dilute pee of the night before.

People who lack functioning aquaporin 2 channels produce large amounts of dilute urine – as much as 25 litres a day – and quickly become seriously dehydrated and very thirsty. This can happen because of a rare genetic mutation, in which case the disease manifests at birth; it can be hard for parents to spot, for urine-soaked nappies are far from uncommon in babies.

Lethal Agents

Ion channels are not only crucial at the start of life – they are also intimately involved in its end. Many cells and organisms use ion channels as offensive weapons. These act as molecular hole-punches, inserting themselves into the membrane of the target cell and forming a huge hole – a giant pore so big that not only ions but also small molecules and essential nutrients can leave the cell. Water rushes in, causing the cell to swell up so much that it eventually explodes (lyses) and dies. Channels used as lethal agents in this way are particularly interesting as they are packaged within the aggressor cell in an inactive form in which they do no harm. Once released, they reassemble themselves into a structure that is able to embed itself in the membrane of their prey. They are true transformers, shape-shifting from a harmless inactive form to a highly lethal one in matter of seconds.

Such channel-forming molecules play important roles in our immune system, defending us against invading pathogens. One type, appropriately named defensins, is found in our skin and the lining of the airways, where they act as natural antibiotics with a broad spectrum of action against bacteria, fungi and some viruses. Others are released by specialized white blood cells known as killer T-cells (or natural killer cells). Killer T-cells kill viruses and bacteria in a number of different ways, but one of them is by releasing perforins – ion channels that punch holes in alien cell membranes. Another pore-forming weapon in the arsenal of our immune system is complement, which produces even larger perforations in invading cells.

Bacteria also indulge in incessant chemical warfare with one another, secreting channel-forming proteins that kill other bacteria. Unfortunately, some of these also attack human cells. Alpha toxin, secreted by *Staphylococcus aureus*, is one of the largest, most lethal and most beautiful of all. It is a mushroom-shaped channel, with the stalk spanning the membrane and the cap resting on its outer surface, projecting out from the cell. To avoid damaging the bacterium itself the channel is made of seven separate subunits, which are secreted individually and subsequently co-assemble to form a giant pore that punctures the target cell. *Staphylococcus* bacteria cause skin infections such as carbuncles, boils, abscesses, wound infections, and, most seriously of all, systemic infections in which the bloodstream carries the toxin and bacteria to all tissues and both red and white blood cells may be damaged (causing blood poisoning). The ability of alpha toxin to lyse red blood cells gives rise to its alternative name, haemolysin.

Staphylococcus pyrogenes, the bug that causes scarlet fever, also produces a toxin that bursts red blood cells, causing a characteristic fine red rash all over the body and a bright strawberry-red tongue. It can be fatal – the sister of the nineteenth-century American novelist Louisa May Alcott died of the disease, a traumatic event that the writer subsequently used in her novel *Little Women*. Other ion channels, such as those released by the protozoan that causes amoebic dysentery, wreak havoc in our guts.

Battling Bugs

Humans have harnessed such channel-forming bacterial toxins for their own purposes. Some, which attack bacterial cells but not mammalian ones, have been exploited as antibiotics. Others are used as insecticides. The best known is that secreted by the bacterium *Bacillus thurigiensis*, which inserts itself into the cells lining an insect's gut, causing them to lyse, so that the insect eventually dies of dehydration. The toxin is released as an inactive precursor that must be activated in the insect gut and so is harmless to humans.

Bacillus thurigiensis is widely used as a biological control agent to limit caterpillar populations in commercial greenhouses, to destroy mosquito larvae, and to kill the blackflies that carry river blindness. More recently, the gene that codes for the bacterium's toxin has been engineered into plants, which then manufacture the toxin themselves. Pesticide-producing strains of maize, potato and cotton are commonly grown in the USA and enable the use of synthetic insecticides to be dramatically reduced. This has had clear environmental benefits. Nevertheless, the practice has been quite controversial, in part because of anxiety about genetically modified crops. Another concern is that continual exposure of insects to the pesticide creates a strong evolutionary selection pressure that favours toxin-resistant insects. Any insect developing a mutant receptor that does not bind the toxin has a clear reproductive advantage, and insects resistant to the pesticide have already been reported. As is the case for antibiotics, countering resistance is a constant battle.

Cell Suicide

Long ago, before you were born, you had webbed hands and feet like those of a duck. As you developed inside your mother's womb, the cells that made up the web of soft tissue between your digits were killed off in a process known as programmed cell death (or apoptosis) so that you ended up with separate fingers and toes. If this process of body sculpting fails, as occasionally happens, you end up born with webbed fingers.

Everyone who has kept tadpoles has seen such cell suicide in action for the gradual disappearance of the tadpole's tail as it develops into a baby frog occurs by apoptosis and reabsorption of the dying cells. Similarly, apoptosis is drawn to the attention of a woman every month, for the sloughing off of the lining of the womb that occurs at the start of her period is also the result of programmed cell death. Perhaps most important of all, cell suicide plays a key role in the development of the nervous system and in how your brain is wired up. Early in development, many nerve cells are born and send forth their axons towards their destination in an exploratory manner. If they find their correct targets, a tentative connection is established, impulses speed excitedly down the lines, chemical kisses are exchanged, and the link is cemented. Nerve cells whose axons fail to find their correct targets produce more feeble impulse activity and simply wither away through lack of use. Many die during brain development and without such cell suicide the brain could not function correctly. Apoptosis is also a way to ensure that damaged cells that might threaten an organism's survival are eliminated. Your immune system can kill cells infected with viruses this way, and cells whose DNA is damaged are encouraged to commit suicide to prevent cancers forming.

At the cellular level, then, death is far from being a negative event. It is an essential part of the life of every multicellular organism and every day several billion cells in our bodies die by apoptosis. Without it, multicellular life is not possible. If we are no closer to understanding the meaning of life, at the cellular level, at least, we might claim to understand the meaning of death.

A Time to Live, a Time to Die

When a cell commits suicide it shrinks, its membrane lifting away from the underlying cytoplasm in ugly bubble-like blebs. The DNA is broken down so that no more proteins can be produced, and the mitochondria, the cell's powerhouses, are disabled. Specific lipids appear on the surface of the cell membrane that signal to scavenger cells to come and gobble up the broken fragments of the dying cell for recycling.

There are several ways in which a cell can self-destruct but, as you have probably guessed, one of them is mediated by an ion channel. It also involves the mitochondria, tiny intracellular organelles, about the size of a bacterium, that are found in almost every cell of your body. The ancestors of mitochondria were once free-living entities, rather similar to the blue-green algae (the cyanobacteria) that form the familiar green scum on lakes in hot summers, but around two billion years ago these ancestral mitochondria gave up the solitary life and became incorporated within early cells. Thus like the *Star Trek* aliens known as the Trill, we live our lives in partnership with another organism – but this is no science fiction and our symbionts are microscopic. Almost all plant and animal cells contain mitochondria and they are essential for life: without them, multicellular organisms could not function, as mitochondria act as molecular furnaces where fuels such as sugar and fats are burned with oxygen to produce chemical energy. Cells that require a lot of energy, like muscle cells, have large numbers of mitochondria.

But mitochondria also have their dark side. They are surrounded by two membranes, whose integrity is important for the ability of the mitochondrion to produce energy. When a cell decides to commit suicide a large pore forms in the outer mitochondrial membrane known as the mitochondrial apoptosis-induced channel. The hole is so big that relatively large chemicals can leak out of the mitochondria into the cytoplasm, where they create mayhem, triggering a cascade of events that leads inexorably to cell death. Importantly, however, the decision to commit suicide is not decided by the mitochondria itself. It is a process that is initiated and tightly controlled by the cell, which simply co-opts the mitochondrial machinery to serve its own ends.

Blighted Harvest

Mitochondria are also targeted by the Southern Corn Leaf Blight toxin, which wreaks such havoc on cytoplasmic male sterile (CMS) strains of maize. CMS maize plants are sterile because they possess a unique ion channel that sits within their inner mitochondrial membrane. Like a silent timebomb, this channel is normally closed and does not affect organelle function. However, binding of the

Southern Corn Leaf Blight toxin activates the timebomb, opening the channel and destroying the ability of the mitochondria to make energy. Without energy, the cell dies. As the fungus spreads, the toxin destroys the plant, cell by cell. Only those plants that possess the ion channel gene, that is the CMS varieties, are susceptible. It is an inescapable association, for toxin sensitivity and male sterility result from the same process. Even in the absence of the toxin, the ion channel is activated in the mitochondria of the cells that supply the developing pollen grains with nutrients, and when these cells wither and die, so too does the pollen.

Despite the wide devastation caused by Southern Corn Leaf Blight in 1970 in the USA, the country was lucky. More than 85 per cent of maize plants at that time carried the gene. Only the dry September in the northern and western states, which limited the spread of the fungus, prevented an almost total destruction of the crop. As Paul Raeburn points out in his thought-provoking book *The Last Harvest*, the size of the Southern Corn Leaf Blight epidemic and its enormous economic impact resulted from the fact that the Corn Belt in the USA was largely planted with a single variety of maize. The genetic uniformity of modern crops and the practice of planting only one or two varieties over a wide area means that if one plant is susceptible to a new disease, all plants will be. Consequently, the whole crop is at risk. More traditional methods of agriculture, which use many different local varieties, preserve considerable genetic variability so that although some plants may succumb to infection, many others will be resistant. A good reason, then, for preserving as many wild crop species and indigenous cultivars as we can, for without the genes that these plants contain, plant breeders may be unable to adapt crops to the new dangers we will assuredly encounter in the future.

Green Electricity

Almost all life on the planet depends on the ability of plants to capture the energy of the sun and store it as sugar molecules. This process, known as photosynthesis, is the ultimate source of all the food we eat, all the molecules from which our bodies are built, and most of the oxygen in the atmosphere. Photosynthesis involves the conversion of carbon dioxide and water into sugar and oxygen in a reaction powered by sunlight, and it takes place in specialized organelles, known as chloroplasts, that lie within plant cells.

To prevent excessive water loss, the leaves of most plants are covered with a thick waxy cuticle. However, this also restricts the diffusion of oxygen and carbon dioxide into and out of the leaf, so that gas exchange can only take place via dedicated pores on the underside of the leaf, known as stomata, that act like microscopic windows. The problem that plants face is that the stomata not only allow carbon dioxide to enter and oxygen to leave, they also provide a very effective pathway for water vapour to escape. This can put a considerable strain on the plant, for water lost this way must be replaced by absorbing more from the ground. Some desert plants have solved the problem by opening their stomata only at night, greatly restricting water loss during the heat of the day. But this poses another difficulty because photosynthesis normally requires carbon dioxide and sunlight to be present at the same time. It's a classic catch-22 situation. Consequently, most plants balance photosynthesis and water stress by opening and closing their stomata throughout the day, as the ambient light and humidity conditions dictate.

Stomata are composed of two 'guard' cells that both form the aperture of the pore and regulate its opening and closing by adjusting the amount of water they contain. When the guard cells are swollen and turgid the pore between them is forced open, whereas when they lose water and become flaccid the pore collapses shut. The water movements that influence guard cell volume, and thereby stomatal opening, are controlled by a combination of pumps and channels. An increase in light intensity causes positively charged hydrogen ions to be pumped out of the cell, creating a negative potential across the cell membrane. In turn, this change in membrane potential opens potassium channels, allowing potassium ions to enter the guard cells. Water follows the potassium ions, so that the guard cells swell by as much as 40 per cent, forcing open the stomatal pore. As long as the potassium channels remain open, the pore remains ajar. However, when light levels fall or the plant experiences water stress the potassium channels close. Consequently, water leaves the cell, the guard cells shrink and the stomatal pore closes.

In a sense then, by controlling the turgidity of the guard cells, plant potassium channels regulate photosynthesis. Arguably, they are some of the most important ion channels on Earth. I find it strangely pleasing that these potassium channels belong to the same superfamily as the ones I am most passionate about. They must stem from a common ancestor that evolved long ago, before the animal and plant kingdoms divided.

Life in the Slow Lane

Remarkably, a few plants not only have ion channels, they also have the ability to generate action potentials. However, the electrical impulses of plants differ from those of nerves in that they are of longer duration, travel more slowly and are carried by different ions. That of the alga *Nitella*, for example, is initiated not by an influx of positively charged sodium ions, but instead by the loss of negatively charged chloride ions from the cell. There is a good reason why this is the case. Unlike animal cells, the cells of most terrestrial plants are not bathed in a salty extracellular fluid. Ions are present at very low levels in plant cell walls and thus an influx of sodium ions would not be a viable means of producing an action potential. Instead, plants must rely on chloride efflux.

Carnivorous plants have exploited action potentials to capture their prey. One of the most fascinating is the Venus flytrap, a favourite of Charles Darwin. This plant, he wrote, 'from the rapidity and force of its movements, is one of the most wonderful in the world'. To cope with the nitrogen-poor soils of the bogs in which it lives, the Venus flytrap supplements its diet by capturing small insects. It attracts them with an enticing 'trap' formed from a modified leaf that consists of two brilliant crimson lobes, like the two halves of a cockleshell, fringed by long pinkish-green hairs. At rest, the trap sits invitingly open. No sooner has an unwary fly landed on its sweet, sticky surface, however, than the two halves snap shut, imprisoning the insect inside. The long hairs at the edge of the lobes interlock tightly together like the teeth of a rat-trap, preventing large insects from escaping. Small insects can squeeze out, presumably because it would not be energetically favourable to process a tiny morsel, but larger insects are slowly digested to provide the nitrogen the plant needs to make its own proteins. About seven days later the trap reopens, releasing the indigestible remains.

As you will know if you have ever tried to swat a fly, insects move fast. Thus, to catch one, the Venus flytrap must move even faster and it has evolved a specialized electrical signalling system that enables it to do so. Each lobe of the trap bears several triangular hairs projecting up from its surface that are exquisitely sensitive to touch. If more than two of these are distorted at roughly the same time – for example, by the movement of an insect – the lobes clap shut faster than the blink of an eye.[2] The hairs possess mechanosensitive ion channels and touching them elicits an action potential that spreads throughout the lobe cells to the centre of the trap. At rest, the lobes of the trap are bowed upwards, but when the electrical signal arrives at the midline of the trap they flip from a convex to a concave shape, forming a pocket that entraps the prey. Precisely how this happens is still debated, but ion channels that trigger ion and water movements that lead to differential swelling and shrinking of the lobe cells, and thus to dramatic changes in pressure across the leaf, have been invoked.

Similar trapping mechanisms are found in other bog and heathland plants, such as sundews, as well as in the wonderfully named waterwheel plant, which sets its snares underwater. Shutting of the waterwheel trap is one of the fastest of plant movements known, taking only 10 to 20 milliseconds, five times faster than the Venus flytrap.

Although plants do not have nerves, a few have specialized conducting pathways that enable electrical impulses to transmit information for some distance. Tap the leaflets of *Mimosa pudica*, the sensitive plant, and the whole leaf folds up, collapsing from its junction at the stem. Specialized cells transmit the signal from the leaf to its base, where ion movements then cause changes in cell volume that result in the collapse of the whole leaf. By contrast, in the Venus flytrap, the action potentials spread in random fashion throughout the leaf, via electrical synapses between adjacent cells, before finally reaching the swelling cells that close the trap. Nevertheless, impressive as it is that plants have action potentials, they propagate far more slowly than those of animals (around 10 metres per second compared to 120 metres per second). Plants, it seems, simply live their lives at a much slower pace.

9

The Doors of Perception

If the doors of perception were cleansed every thing would appear to man as it is, infinite. For man has closed himself up, till he sees all things thro' narrow chinks of his cavern.[1]

William Blake, *The Marriage of Heaven and Hell*

Imagine you are sitting here with me in my garden on a perfect late summer evening, listening to the blackbird's joyous song, and enjoying a glass of wine and the faint heat of the sun on your skin. You raise your glass and admire the pale-golden colour of the liquid and the glint of the crystal in the sunlight, then bend your head, swirl the wine gently around the glass, and appreciate the light aroma of gooseberries, of sunshine locked in alcohol. You sip and savour the cool taste of the wine. As this simple pantomime illustrates, even something as simple as drinking a glass of wine involves all of our senses.

Pleasure, pain, indeed the evolutionary success of any organism, ourselves included, depends on our ability to perceive the world around us: to see, hear, smell, taste and touch it. Our sense organs convert the myriad signals that constantly bombard us in multiple modalities into a single form that the brain can interpret – the electrical energy encoded in our nerve impulses. And in all cases ion channels are needed to transduce sensory information into that electrical signal. Ion channels are truly the doors of perception as everything we sense is detected, transmitted or processed by them. Consequently, defects in ion channel genes produce a variety of human sensory disorders, from hearing loss to colour blindness. This chapter tells some of the remarkable stories of how ion channels determine our ability to perceive what lies around us. It is concerned with Blake's 'narrow chinks' through which we view the world – our sense organs.

Eye Spy

Our eyes are our windows on the world. Open them, and there lies the world in all its richness of form, movement, brightness and colour. As I sit writing these lines, I gaze out on a painted landscape of countless colours: the clear blue sky of an Indian summer, the faded gold of ripe wheat, a vast palette of different greens laced with splashes of brightly coloured flowers. Nothing is still, for the poplar is shaking its leaves in the breeze and the late roses are being tossed around by the wind.

At one level, our eyes operate like a simple camera. They have a clear cornea and a crystalline lens that work together to focus light rays onto a layer of photosensitive cells known as the retina at the back of the eye. They possess an iris that continuously adjusts the amount of light entering the eye. And they have a protective lens cap – the eyelid – that can shut the light out completely when necessary. Unlike most cameras, however, our eyes have a brain attached that processes and interprets the images projected on the retina. Some processing also takes place in the retina itself.

Every second, our eyes handle thousands of images, transforming light signals into upside-down images on the retina, and converting these into nerve impulses that are sent on to the brain for processing. The transparent outer layer of your eye, the cornea, is responsible for about two-thirds of the focusing power of the eye: the remainder is the province of the lens, which is suspended behind the pupil by thousands of fine ligaments. The cornea has a fixed focus, but that of the lens can change, for muscles attached to its edge pull it thicker or thinner as you focus on near and far objects, respectively. As we grow older the elasticity of the lens decreases, making it harder to change its focus, which is why most people over fifty need reading glasses.

The pupil is the aperture through which light passes. It appears black because no light returns through it. The iris – the coloured bit of your eye – contains muscles that adjust the size of the pupil to the intensity of the ambient illumination, dilating it in dim light and shrinking it to a pinpoint in very bright light. The size of the pupil even signals feeling, for it expands in response to fear, pain or if you see something of interest – someone you love, perhaps.

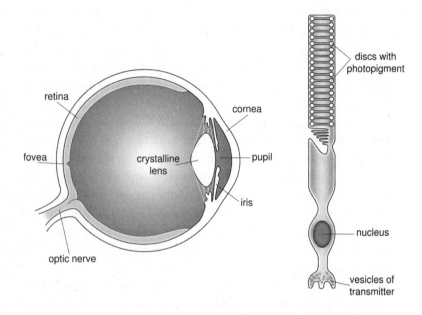

Left. A cross-sectional view of the eye, showing the positions of the cornea, lens and retina. *Right.* A single rod photoreceptor. The outer segment is crammed with stacks of membrane discs that are densely packed with the visual pigment rhodopsin. The other end of the cell is packed with vesicles containing a neurotransmitter. Chemical and electrical signals transmit the light stimulus captured by the photopigment from the discs to the rod terminal and from there on to the next cell in the chain.

The retina is packed with light-sensitive cells, which come in two different varieties: the rods and the cones. Together, they enable us to detect the two fundamental properties of light: intensity and wavelength (colour). Rods cannot discriminate colour, but they are exquisitely sensitive to low light intensities and can even detect a single photon of light (a single quantum or light particle). In dim light we see entirely with our rods, which is why the world appears in shades of silver and grey in starlight and moonlight. Across most of the retina, rods markedly outnumber the cones: there are around 120 million of them against 6.5 million cones. The exception is at the fovea, the region of the retina where light rays are most accurately focused and cones are far more dense. Cones are therefore responsible for our visual acuity, as well as our colour vision. They work best in bright light, which explains why in the dark it is often easier to see things out of the corners of your eyes, where more rods are found. As you will have found, a faint star is far brighter if you do not focus on it directly. The only part of your retina where there are neither rods nor cones is where the optic nerve leaves the eye: this is known as the blind spot, as in the absence of light-sensitive cells nothing can be detected.

Photodetection

The essential feature of any eye, the ability to detect light, is due to dedicated molecules that convert light into chemical energy. We have several such photopigments in our eyes, each of which is specialized to capture different wavelengths (colours) of light. All contain a derivative of vitamin A, retinal, attached to a protein called opsin. The retinal part of the molecule is responsible for absorbing light, which explains why lack of vitamin A decreases the sensitivity of the eye to light and can lead to night blindness.[2] During World War II, the British government spread a rumour that their fighter pilots had been fed extra rations of carrots, which contain a lot of vitamin A, to explain their increased success rate at shooting down enemy bombers. There was no substance in the story, however; it was merely a disguise to cover the introduction of radar, which was in fact responsible for the enhanced 'hit' rate.

The opsin component of the visual pigment tunes the spectral sensitivity of retinal. Thus, by using different opsins, different wavelengths of light can be detected. The photopigment in your rods, known as rhodopsin, is most sensitive to blue–green light, which has a wavelength of 498 nanometres. We have three different types of cones in our eyes, each of which contains a unique photopigment that absorbs maximally at a different wavelength of light. The conventional, but inaccurate, shorthand has been to call these red, green and blue cones, but in fact they detect yellow–green (long, 564 nanometres), green (medium, 535 nanometres) and blue–violet (short, 433 nanometres) light. All of the myriad different hues we can distinguish are created by combining the electrical signals from these three types of cone, just as a colour television uses only three colour signals to produce the many different colours we see on the screen.

Whenever one of the photopigments captures a photon its shape changes. This sets in motion a complex cascade of events that eventually results in a change in the electrical properties of the light-sensitive cells. Our ability to see begins with this transduction of light into an electrical signal, in which ion channels play an intimate role. Both rods and cones contain millions of pigment molecules, crammed together within the membranes of a series of intracellular discs that are stacked up in the outer segment of the cell. The enormous number of molecules greatly increases the chance that a photon travelling through the eye will be captured, and trigger a visual response. But the location of the photopigment presents a problem, for it lies far away from the vesicles containing the transmitter that the photosensitive cell uses to signal to its neighbour. Thus photoreceptors employ an intracellular messenger service to link the light-induced conformational change in the photopigment to transmitter release. This consists of a chemical known as cyclic GMP, which relays information from the intracellular discs to the surface membrane of the cell. Here, the chemical signal is converted into a much faster electrical one that is rapidly conducted to the transmitter release sites. At the heart of this complex signalling cascade is a special kind of ion channel that opens when it binds cyclic GMP.

In the dark, cyclic GMP levels in the rods and cones are high, so that the cyclic GMP-gated channel is held open. Sodium ions flooding in through the channel pore produce a positive swing in the membrane potential that spreads over the surface membrane to the other end of the rod or cone cell. There, it stimulates calcium channels to open, enabling calcium ions to enter the cell and trigger the release of a transmitter that stimulates the next cell in the chain and thereby tells the brain that it is dark.

Light-induced changes in the visual pigments activate a signalling cascade that leads to the destruction of cyclic GMP. As a consequence, the cyclic GMP-gated channel closes, switching off transmitter release and signalling 'Light!' The exquisite sensitivity of our vision derives from this complicated chain reaction, which constitutes a powerful amplification system. Many cyclic GMP molecules are destroyed for every photon captured, ensuring that enough channels close to switch off all transmitter release. As you will already have appreciated, the other remarkable thing about our rods is that they signal continuously when they are not being stimulated, being active in the dark and switched off by light. This feature is also thought to enhance our sensitivity to light.

Viagra (sildenafil) is widely used to counter impotence and enhance sexual performance, as the many junk emails I receive testify. But it also has a lesser-known effect. At high concentrations, it can literally turn your world blue. Men taking high doses of the drug sometimes find that it produces a transient mild blue tinge to their vision and an increased sensitivity to light. This is because Viagra has a weak inhibitory effect on the activity of the enzyme that destroys cyclic GMP in rod cells and so enhances their sensitivity to light. Because of the possibility that Viagra can interfere with colour vision, and impair the ability to distinguish blue and green lights on airport taxiways, the Federal Aviation Administration prohibits pilots from flying within six hours of using the drug.

Seeing in the Dark

In daylight we use our cones to see with, as our rods do not work. This is because bright light bleaches the rhodopsin molecule so that it no longer responds to light. It takes time to regenerate, as you can easily discover for yourself if you walk straight from a brightly lit environment into a blacked-out room. At first, you will see nothing, as your rods are still inactivated by their previous exposure to high light intensity. Gradually, however, you will grow accustomed to the dark, and faint shapes will slowly start to appear from the shadows, growing ever more solid with time as your rhodopsin recovers from bleaching. It takes about thirty minutes to fully regenerate all your rhodopsin – and it will be lost within seconds if you step out into the light again.

I vividly remember one occasion when a colleague was trying to record electrical signals from a frog's eye, at the Cold Spring Harbor Laboratory in New York. To do so, he needed the rods to be dark-adapted, so he put the frog in a dark room. When he returned, thirty minutes later, the frog had escaped. I offered to help. This was a mistake because, not wishing to delay his experiment, my colleague refused to turn on the light. Capturing an energetic frog in a small, cramped, dark room illuminated only by the feeble red beam of a tiny pocket torch, while trying to avoid bumping into an increasingly frustrated and wild-eyed companion was a surreal experience. Because rods are insensitive to red lights, and so are not bleached by them, we used a red torch for this escapade, relying entirely on our cones for vision. For the same reason, red lights are used for the instrument displays of ships and airplanes at night, where it is important that pilots can read them without losing their night vision.

Seeing Red

Colour, like beauty, is in the eye of the beholder. It is not a property of light itself, as Thomas Young first recognized back in the early nineteenth century when he proposed that colour sensation was encoded by three different kinds of pigment. Colour is actually constructed by a collaboration between the eye and brain of the observer, but Young was correct about the three cone pigments. It is thought that humans may have evolved colour vision to enable us to see ripe, orange–yellow fruit on green trees and the yellowish colour of juicy young shoots, for which we need three types of cones. Most mammals, such as dogs and cats, have only two kinds of cone photopigment and so see only a limited range of colour: contrary to popular belief, bulls do not see red. Other creatures live in a world entirely without colour. But humans should not be too complacent, for we are far from having the best colour vision in the animal world and lag well behind the mantis shrimp, which enjoys ten or more different visual pigments. Even tropical fish possess four or five types of cone.

We can see light at wavelengths between roughly 400 and 700 nanometres, which corresponds to the blue and red ends of the visible light spectrum. Other creatures once more surpass us for they may see wavelengths far beyond this. Pit vipers, vampire bats and fire beetles, for example, sense infrared using specialized organs to detect heat. Most birds and insects possess an additional photopigment that enables them to detect ultraviolet light and flowers have evolved ultraviolet markings on their petals to guide butterflies and bees to their nectar stores. Male and female blue tits look similar to us, but not to their mates, for they carry bright flashes of ultraviolet reflective feathers on their crests. Bizarrely, urine stands out clearly in ultraviolet light, a fact exploited by birds of prey who track small rodents by the urine trails they use to mark their territory. Reindeer are also sensitive to near ultraviolet, which is thought to help them find food in a white-out, as in ultraviolet light the pale lichens on which they feed stand out starkly black against the white snow.

Through a Lens, Darkly

Even the colours you do see can be distorted. The lens of the eye starts out as crystal clear, but over time the transparent proteins of which it is composed may become damaged by continual exposure to ultraviolet light and clump together, so that the lens progressively becomes opaque and takes on a yellowish tint. As the cataract develops, the world gradually becomes blurred and hazy and its colours change: white becomes a dull yellow, greens become yellows, bright reds mutate into muddy pinks, blues and purples morph into red and yellow. These colour changes are very evident in Claude Monet's later paintings. Soon after he was seventy, he began to develop cataracts in both his eyes and the lovely, delicate impressionistic nature of his painting may derive in part from the fact that he increasingly saw the world as blurred. But he was very frustrated that his poor eyesight meant he could no longer see colours with the same intensity he remembered and he was forced to carefully order the sequence of paints on his palette to aid their identification. After 1915, the emphasis on red and yellow colours in his work became particularly pronounced and the light-blue shades disappeared. He had particular difficulties with some of the Water Lily paintings he was working on, and having decided that he was no longer capable of painting anything beautiful, he destroyed several canvases. Finally, when he was in his early eighties and almost completely blind, he underwent surgery to remove the cataract in his right eye. Initially he was bitterly disappointed with the result, complaining of the changes in colour he perceived, but after a second operation to remove the cataract in his other eye he regained confidence and produced his wonderful late Water Lily canvases. These resemble his earlier paintings more closely than the ones he produced while suffering from cataracts.

Every year around 120,000 people in the UK have a cataract removed. Indeed, many of us will have this operation if we live long enough, for cataracts are a common side-effect of ageing. The operation is simple and transformative. Afterwards the world suddenly snaps into sharp focus and colours appear crystal clear. As my mother remarked, 'the dirty yellowish shirts that I could never seem to wash clean were suddenly revealed to be a bright pristine white – it was like a washing powder advertisement'. Some people may also see the world in an entirely new light. The lens of our eye not only serves to focus light rays, it also cuts out ultraviolet light. Most people have an artificial lens implanted when their own is surgically removed in a cataract operation. But not all. Those who do not then see the world through new eyes for, like bees and butterflies, they become sensitive to ultraviolet light and everything appears brighter and bluer.[3] The fact that Monet did not receive a new lens when his cataract was removed may have influenced the mauve and violet hues in his later paintings.

Extraordinary Facts Relating to the Vision of Colours

In 1798 the chemist John Dalton described his own colour blindness in a lecture held in Manchester, and published the first scientific account of the condition in an accompanying paper. In Dalton's view, the colour of grass closely matched the paint-box red of sealing wax, which led him to conclude that he saw either red or green differently from other people. He also found it difficult to distinguish blue and pink, and to his considerable surprise, many colours appeared different by candlelight. He wrote that he 'was never convinced of a peculiarity in my vision, till I accidentally observed the colour of the flower of the Geranium zonale by candle-light, in the Autumn of 1792. The flower was pink, but it appeared to me an almost exact sky-blue by day; in candlelight, however, it was astonishingly changed, not then having any blue in it, but being what I call red, a colour which forms a striking contrast to blue'.

Dalton interpreted these findings to indicate that the fluid in the cavity of his eye was tinted blue and so selectively absorbed the longer wavelengths of light, and he instructed that his eyes should be dissected after his death to see if this was the case. The gruesome experiment was duly performed the day after he died but the fluid was totally translucent. Two hundred years later, modern DNA technology was used to identify the cause of Dalton's colour blindness, using fragments of Dalton's eyes that had been carefully preserved by the Manchester Literary and Philosophical Society. It turned out he had an inherited condition known as deuteranopia.

Many men have impaired colour vision as a result of mutations in one of the three different kinds of cone visual pigments that detect yellow–green, green or blue–violet wavelengths of light. Most commonly, the yellow–green and green photopigments are affected. About 2 per cent of men completely lack one pigment, giving rise to a condition known as protanopia (absence of yellow–green photopigment) or deuteranopia (no green photopigment), and in 6 per cent of men the spectrum of one photopigment is shifted, so that colour is seen differently. In all cases, it is hard to distinguish between red and green, which both appear a muddy yellowish-brown. It is always worth remembering when preparing colour slides for a presentation that some of your audience may not be able to distinguish red and green colours easily. Similarly, it can be hard for the colour blind to distinguish between ripe and unripe fruit, and some foods appear the unappetizing colour of excrement. The genes for the yellow-green and green visual pigments are found on the X-chromosome, of which men have only a single copy, which explains why far more men than women are red–green colour blind. In women, the gene on the other X-chromosome can substitute for the defective one.

Some people, known as achromatopes, are born with a rare genetic condition that means they have only rod vision and cannot see colour at all. Such total colour blindness is very rare. In the general population, the incidence is about 1 in 30,000 people but as Oliver Sacks relates in his book *The Island of the Colour-blind* it is far more common on the Micronesian island of Pingelap, where 5 per cent of people are affected. It is thought they are all descended from a single individual, a carrier of the mutant gene, who was one of only twenty survivors of a typhoon that struck the island in the 1770s. Achromatopsia is due to a total lack of functioning cone cells. One of its main causes is a mutation in the cyclic GMP-gated channel of the cone cell, as is the case for the Pingelap islanders. Because a different gene codes for the rod channel, people with such mutations are not totally blind. They are, however, dazzled by bright lights and they find it hard to see in normal daylight because their rods cease to function at high light intensities. Even people with 'normal' colour vision may not see the world in exactly the same way. Variants in the DNA that codes for the visual pigments may produce subtle differences in our perception of colour. The red that I see may not be quite the same as that which you do.

On 15 November 1875, there was a terrible train crash at Lagerlunda in Sweden when two express trains travelling on the same single line track collided head-on. The driver of the late-running northbound train appeared to have ignored the red light the stationmaster had been waving at him to signal him to stop, and because he had died in the crash it was not possible to question him. The physiologist Professor Alarik Frithiof Holmgren investigated and concluded that the accident had been caused by the colour blindness of the driver.[4] As a consequence, Sweden implemented a colour blindness test based on the ability to discriminate different coloured wools and other countries followed suit shortly afterwards. Today, people with colour blindness are banned from certain professions, including that of an airline pilot, and until very recently a few countries, like Romania, even denied people who are colour blind a driving licence.

Hear, Hear!

Our world is full of sounds. A Bach cantata, the roar of traffic, the swoosh of the seashore, the rustles of leaves, the chatter of children, the low throb of a generator, the high-pitched scream of a swift – all travel to our ears as pressure waves and are effortlessly converted by our brain into the sounds we hear. This process is extraordinarily sensitive for we can hear sounds quieter than the tinkle of a pin dropping and discriminate those that are separated by just one-thirtieth of a semi-tone, the smallest interval in Western music. How is it possible that we can distinguish such a range of sounds or pick out one soft voice against a background roar?

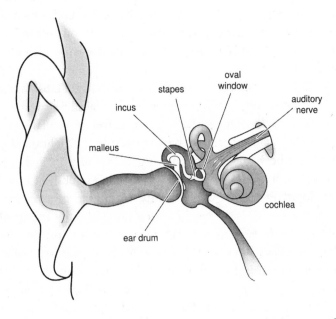

The outer and inner chambers of the ear, showing how displacement of the ear drum is transmitted via the three ear bones (malleus, incus and stapes) to the fluid-filled cochlea where the hair cells transduce sound vibrations into electrical impulses.

An elegant description of how we hear is given by Aldous Huxley in his novel *Point Counter Point*. 'Pongileoni's bowing and the scraping of the anonymous fiddlers had shaken the air in the great hall, had set the glass of the windows looking onto it vibrating; and this in turn had shaken the air in Lord Edward's apartment on the further side. The shaking air rattled Lord Edwards' membrana tympani; the interlocked malleus, incus and stirrup bones were set in motion so as to agitate the membrane of the oval window and raise an infinitesimal storm in the fluid of the labyrinth. The hairy endings of the auditory nerve shuddered like weeds in a rough sea; a vast number of obscure miracles were performed in the brain, and Lord Edwards ecstatically whispered "Bach!"'

As Huxley relates, sounds are simply pressure waves in the air that radiate outward from a sound source, much like the ripples in a pond.[5] These are collected and filtered by our ears and funnelled towards the eardrum, which vibrates in response. In turn, this moves three delicate interlocking bones, the malleus ('hammer'), incus ('anvil') and stapes ('stirrup'), which are among the smallest bones in our body and no bigger than the size of a single letter of this print. These relay the vibrations to another membrane, the oval window. At this point the sound waves pass from air to the fluid-filled canals of the inner ear, where sensory cells convert the sound waves into electrical impulses. These are then forwarded via the auditory nerves to the brain, where they are interpreted.

The ear must register both the intensity and frequency (tone) of a sound. Nerve cells are not especially well suited to this, for their maximum rate of firing and the range of intensities they can signal are quite low. Yet the loudest sound we can hear may be 100,000 times greater than the quietest, and we can detect tones ranging from frequencies as low as 20 hertz (20 cycles per second) to as high as 20,000 hertz. So how do our ears do it?

Making Waves

The most important part of the ear – the bit that actually senses sounds – is tucked safely away inside our skull. This is the cochlea, a fluid-filled sac that is coiled up tightly like a snail shell to fit inside the temporal bone (hence its name – cochlea is Latin for 'snail'). About the size of a pea when coiled up, it is 35 millimetres long when unrolled and split longitudinally into three compartments by two membranes. Around 16,000 specialized sensory cells called hair cells are arranged along the length of the lower (basilar) membrane in four rows: three rows of outer hair cells and one of inner hair cells. On their upper surface the hair cells bear bundles of stiff eponymous hairs, known as stereocilia, that reach up toward the tectorial membrane.

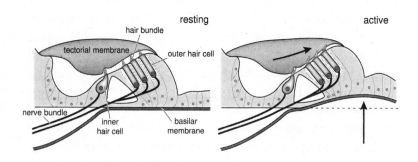

The region of the cochlear that enables us to hear, shown in the resting state (*left*) and in the active state (*right*). Displacement of the basilar membrane (*right*) causes the hairs on the tip of the hair cells to be displaced. In the case of the inner hair cells, this causes the sensation of sound.

Sound waves induce oscillations in the cochlear fluid on either side of the basilar membrane, causing it to vibrate. By experimenting on human cadavers, a Hungarian engineer, Georg von Békésy, showed that sound moves along the basilar membrane as a travelling wave (like that seen when a whip is cracked), building up in amplitude to reach a peak at a particular point along the length of the membrane and then rapidly dissipating. Where this peak occurs depends on the frequency of the sound: high-frequency sounds move the base of the basilar membrane most, and low frequency sounds produce the greatest deflections at the tip of the cochlea. The tiny movement of the basilar membrane is transferred to the hair cells, causing their stereocilia to be displaced backwards and forwards, so producing a mechanical deflection that opens specialized ion channels.

These mechanically gated ion channels are at the heart of hearing for it is they that convert sound waves into electricity – or more precisely mechanical energy into electrical energy. The molecular identity of the channel is not yet known, but the way in which they are opened has been elucidated and it is remarkable. The stereocilia the hair cells bear are organized in rows of decreasing height and are connected to one another at their tips by a tiny thin stiff rod known as a 'tip link'. One end of the tip link is also connected to the mechanosensitive channels that sit at the tip of the stereocilia. As the basilar membrane moves up and down, the tip links are stretched or compressed, pulling open, or forcing shut, the ion channels, respectively. When they open, positively charged ions rush into the cell and alter the voltage gradient across the hair cell membrane. This electrical change has different effects depending on whether the cell is an inner or an outer hair cell.

Picking up Good Vibrations

The inner hair cells are responsible for converting sound waves into electrical impulses and forwarding them to the auditory nerve. The change in voltage across the inner hair cell membrane produced by sound of a specific frequency triggers the release of a chemical transmitter. This stimulates impulses in the terminals of the auditory nerve and so sends signals to your brain. Inner hair cells at different positions along the basilar membrane respond to different frequencies, with those at the base of the cochlea detecting high-pitched sounds and those at the tip responding to low-pitched notes. This frequency discrimination simply reflects the magnitude of the movement of the basilar membrane – recall that high-pitched sounds have their greatest effect at the base of the cochlea. Nerve fibres coming from different regions of the basilar membrane are therefore tuned to specific frequencies, enabling the brain to discriminate different tones on the basis of which fibres are active. This complex bit of molecular machinery is in action inside your head right now, as you listen to the sounds around you.

Dancing Hair Cells

The outer hair cells are far more numerous than the inner ones. Although they play little, if any, role in signalling sounds to the brain, they are essential for normal hearing as they mechanically amplify sound vibrations by 'dancing' in time to the beat. This amplification is critical for detecting low-intensity, high-frequency noises because the vibrations of the sound waves are damped down as they pass through the fluid-filled canals of the inner ear.[6] Without magnification of the signal, the inner hair cells are not stimulated sufficiently to activate the auditory nerves. The cochlear amplifier, as it is known, also sharpens the ability of the ear to discriminate different frequencies. It may have evolved to enable the first mammals to hear the faint high-pitched cries of their young; now it helps us hear the squeak of a bat.

The existence of a natural amplifier in the ear was first proposed in 1948, but the idea was rejected and it was not until the late 1970s that its validity was established. The fact that hair cells could be seen to 'dance' provided the clue to how it worked. An outer hair cell will twitch in time to music piped directly into the cell via an electrode connected to an amplifier. I've not forgotten the time I visited Jonathan Ashmore's laboratory at University College London and looking down the microscope was amazed to see a tiny hair cell bopping away to 'Rock around the Clock'. It kept perfect time. The contractions of the hair cell are powered by prestin, a molecular motor that is sensitive to the voltage difference across the cell membrane, and changes in this voltage difference produced by the exogenous electrical stimulus caused the cell to dance. In life, such voltage changes are produced by opening of mechanosensitive ion channels in response to the movement of the hair bundle within, as

Huxley so evocatively put it, the storm in the cochlear fluid. The twitching of the outer hair cells amplifies the movement of the basilar membrane, leading to greater stimulation of the sensory inner hair cells. This intrinsic biological amplifier is the basis of our ability to hear very quiet sounds. High doses of aspirin inhibit the motor protein and induce a reversible hearing loss.

The Song of the Ear

It may come as a surprise, but your ears can also generate sounds. Technically known as otoacoustic emissions, these are produced by the outer hair cells. They arise because the vibrations the hair cells produce as they bounce up and down set up waves in the cochlear fluid, which in turn are passed on to the air in the middle ear and ultimately back to the eardrum. The sounds made by healthy ears are quieter than a whisper and those of individuals with cochlear damage are even weaker. However, they can be picked up by a special microphone placed within the ear canal, and such 'ear songs' provide doctors with a valuable non-invasive measure of the health of the ear and a simple way to check if a young baby has impaired hearing. This enables a child to be fitted with a hearing aid or cochlear implant before the time window for learning speech has passed.

Living Under a Deaf Sentence

Helen Keller, who was both blind and deaf, once said that whereas blindness separates people from things, deafness separates people from people. The isolation, confusion, frustration and depression often experienced by those who lose their hearing is poignantly stated by Ludwig van Beethoven in his 'Heiligenstadt Testament', written at the age of thirty-two, six years after he had started to go deaf. 'Oh you men who think or say that I am malevolent, stubborn or misanthropic, how greatly do you wrong me. You do not know the secret cause which makes me seem that way to you [. . .] it was impossible for me to say to people, "Speak louder, shout, for I am deaf." [. . .] for me there can be no relaxation with my fellow men, no refined conversations, no mutual exchange of ideas. I must live almost alone, like someone who has been banished.' By the time he was forty-five, Beethoven was almost totally deaf. Yet although performance became impossible, he continued to compose and conduct. At the première of his Ninth Symphony (when he was fifty-four), he had to be gently turned around to witness the rapturous applause of the audience, for he could hear nothing. He wept at the sight.

Unlike a musician, a deaf painter is still able to practise his art. Indeed, in Goya's case it seems to have led to his greatest works. After a serious illness left him stone deaf, the devastating isolation he experienced precipitated a remarkable change in his art: increasingly, he focused on nightmarish fantasies, black visions and satirical portrayals of human behaviour. Liberated from the cacophonous distractions of daily life, it is said, he saw the world for what it was – although whether Goya, like the art critics, perceived his deafness as a blessing is questionable.

Hear Today, Gone Tomorrow

It is estimated that about nine million people in the UK – as many as one in seven of the population – suffer from some degree of hearing loss. Almost inevitably, it seems, as we grow older our ability to hear high frequencies declines. Alan Bennett wrote, 'I did not think my hearing had deteriorated at all but [. . .] R. asks me if I can hear the crickets and I cannot believe that, the night tingling with the sound, I am dead to it.' Such high frequency hearing loss often creeps up on you gradually, almost unnoticed, until suddenly hearing is gone. The poet Philip Larkin only discovered it when a friend remarked on the beauty of a skylark's song and he was unable to hear it. For him, as for many others, partial deafness darkened his melancholy. Age-related high frequency hearing loss has even been exploited by the manufacturers of a controversial device termed the Mosquito that emits an ear-splitting high-pitched whine, audible to teenagers but not to adults. Its painful buzz has been used to disperse loiterers and prevent antisocial behaviour on UK streets. In an ironic twist to the story, ingenious teenagers subsequently stole the frequency and used it as a mobile phone ringtone that they – but not their teachers – could hear.

The deterioration in hearing we all experience with age is caused because our hair cells naturally die off with time and once gone they are lost forever. Loud noises destroy our hearing even faster. The rock musician Pete Townshend[7] instantly lost the hearing in one ear in a famous incident in which a staged explosion was far louder than expected. Thousands of troops fighting in Iraq and Afghanistan returned home with permanent hearing loss, mostly caused by roadside bombs. The thunder of warfare, the deafening sound of a pop concert, the roar of jets and loud machinery – all exact a heavy toll. This is because our outer hair cells are very vulnerable and can be irretrievably damaged by loud noises.

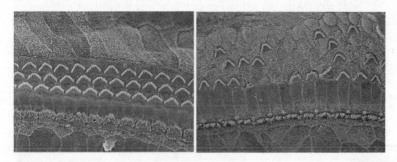

Left. Normal hair cells, showing the three layers of outer hair cells and, below, the single layer of inner hair cells. *Right*. Loud noises damage the outer hair cells before the inner ones.

Chronic exposure to moderately loud sounds can also cause permanent hearing loss because there is no time for partially damaged hair cells to recover. Many people, often unwittingly, routinely subject their ears to noise levels that can ruin their hearing. Exposure to sounds louder than 85 decibels for an extended period of time can cause hearing loss: this noise level is similar to that associated with using a power drill, riding a motorcycle, going to the cinema and many other everyday pursuits. It is also lower than the maximum volume levels on many portable MP3 players. Turn up the volume too loud for too long and you may be unable hear your grandchildren in later life. Sadly, it seems inevitable that within the next few decades many people will become far more interested in how their ears work than they might wish.

One of the first signs of damage is a chronic ringing in the ears known as tinnitus. *The Times'* music critic Richard Morrisson, who tested a device that simulates tinnitus, described it as a horrible, high-pitched whistling that made listening to music a nightmare. 'It was like listening to the drifting signal of some Algerian radio station through the crackling static of an old wireless. Only much more distressing.' Morrisson found immediate relief by ripping the simulator off his ears but for those unfortunates whose hair cells are ruined tinnitus can be a lifelong ordeal. For them, silence is never silent. Beethoven, who suffered from severe tinnitus from his late twenties, complained that his ears whistled and buzzed continually, day and night, and described his condition as truly frightful. It is extraordinary that despite this handicap he was able to compose some of the world's greatest music.

Although tinnitus is often associated with hearing loss, this is not always the case and many tinnitus sufferers hear perfectly well. What causes these internally generated sounds to be perceived is still far from clear, but we do know that they originate from changes that take place within the brain.

A Matter of Taste

I first tasted the miracle fruit one hot summer's afternoon in Puerto Rico. This smooth oval-shaped red fruit, about the size of a coffee bean, comes from the shrub *Synsepalum dulcificum*, a native of West

Africa, and has the extraordinary property of making sour things taste sweet. It felt cold and hard as I rolled it over my tongue and I bit into it with a mixture of anticipation and trepidation. It had a thin, bitter skin surrounding yellow, slightly astringent flesh and a quite unremarkable taste. Ten minutes later I was able to eat a lemon without wincing and, somewhat tentatively, sip vinegar. With my eyes closed, many foods were barely recognizable; beer, in particular, tasted most peculiar. Happily, the effect wore off within a couple of hours.

The miracle fruit contains a protein called miraculin that interacts with sweet taste receptors and enables them to be activated by sour chemicals. Other natural modifiers of taste are also known. If you have ever eaten a fresh globe artichoke you will be aware that everything, including water, tastes sweet afterwards.[8] This is because artichokes contain cynarin, which appears to work by suppressing the activity of bitter taste receptors while enhancing that of sweet ones. Whatever the mechanism, it makes it notoriously difficult to choose a wine to drink with globe artichokes. In contrast, gymnemic acid, from the south Asian herb *Gymnema sylvestre*, suppresses the intensity of sweet perception, but not that of bitter, so that many foods taste unusually bitter and sugar tastes of ashes.

Taste cells are not nerve cells but a specialized kind of epithelial cell (the cells that line the gut, mouth and nasal passages). They are very short-lived, being continually replaced every couple of weeks, and they are packed together in barrel-shaped taste buds. Humans have about 10,000 taste buds distributed over the surface of the tongue, each containing 50 to 100 taste cells.[9] Each taste cell sends a long finger-like process, tipped with fine hairs that bear the taste receptors, up to the opening of the taste bud on the surface of the tongue where stimuli are received. The other end of the taste cell contacts the sensory nerve.

We can discriminate five basic tastes – sweet, salt, sour, bitter and savoury (umami). All the many different flavours we taste, however, are really smelt, for these two senses work in combination. This explains why your sense of taste seems impaired when you have a head cold and your nose is blocked. Anthelme Brillat-Savarin, the seventeenth-century gastronome, tells of meeting a man whose tongue had been cut out, yet who retained a full appreciation of tastes and flavours. He therefore concluded that, 'smell and taste are in fact but a single composite sense, whose laboratory is the mouth and its chimney the nose'.

When you eat something, chemicals contained in the food dissolve in your saliva. This enables them to bind to the receptors at the tip of the taste cells, and so trigger a cascade of events that ultimately releases a chemical transmitter from the base of the taste cell. In turn, this excites the sensory nerve, and nerve impulses are then transmitted to the brain where the information is decoded, processed and tastes are identified.

Different tastes arise because different types of receptor are stimulated. Two tastes – salt and sour – are directly detected by ion channels sensitive to the ions involved, which are respectively sodium ions and hydrogen ions (protons). Salty tastes are mediated by the epithelial sodium channel (ENaC) we met in the previous chapter. Several kinds of ion channel that are sensitive to protons detect sour tastes. The carbon dioxide in fizzy drinks and champagne is also detected by sour taste receptors because it yields protons when dissolved in water. Interestingly, some soda-water manufacturers recognized this long before science showed it to be true – sauerwasser[10] and similar seltzers are named for their sour, slightly acidic taste. Umami, from the Japanese word umai, meaning 'delicious', describes the savoury taste of food containing monosodium glutamate. Some of the receptors that detect glutamate are also ion channels. Somewhat surprisingly, the giant panda lacks functional umami receptors, but whether this is the cause or the consequence of the fact that, unlike other bears, it prefers a strict vegetarian diet is unclear.

Sweet and bitter substances do not activate ion channels directly. Instead they bind to specific receptors, so setting in train a cascade of biochemical events that eventually leads to the opening of a specialized ion channel (known as TRPM5) that is common to both pathways. The ability to discriminate between sweet and bitter substances arises because the two types of receptors are found in different populations of taste cells, which signal separately to the brain. Thus whether something tastes sweet or bitter is decided by the brain. We have over twenty different receptors for bitter taste, but only one for sweet taste, reflecting the evolutionary drive to identify bitter-tasting substances, which are often poisonous. The sweet taste receptor is composed of two different proteins and variants in either of the genes that encode these proteins give rise to different sensitivities to sweet substances; it seems that some people really do have more of a 'sweet tooth' than others. Reduced sensitivity to sugar is most common in sub-Saharan African populations, suggesting that the ability to sense sugar is more important in cold climates, where sugar sources are rare. But in today's society the beguiling pleasure of sweet taste brings in its wake terrible public health problems – obesity and tooth decay are the handmaidens of Sachertorte, raspberry ice cream and sugary drinks.

Many patients taking anti-cancer drugs complain that food tastes terrible – less sweet and more bitter. This is because, like all epithelial cells, taste cells have a very rapid turnover and thus are especially sensitive to chemotherapeutic drugs, which destroy rapidly dividing cells. Taste is also influenced by context (although this is largely the province of the brain). I love the smell of coffee but gave up drinking it over twenty years ago and now take only tea. On the odd occasion when I am accidentally handed the wrong drink and take a sip of coffee it tastes very strange. The ability to identify the correct flavour is also reduced if the food is the wrong colour; raspberry juice does not taste quite right if it is coloured orange or green. Try it, and see if you agree.

Making Sense of Scents

Scents, as Marcel Proust famously observed, can evoke remembrance of things past. The spicy, peppery smell of lupins reminds me of my great-aunt's garden, crammed with colourful flowers and butterflies, and humming with bees. That of mown hay evokes other childhood memories – of lying in the grass watching the village cricket match, hearing the distant cuckoo and the strangely comforting thwack of leather on willow.

The cells that detect smells lie high up in the nose, almost seven centimetres away from the nostril. These are the olfactory neurones, which send processes to the olfactory epithelium in the nose. Each nerve process terminates in a small bunch of olfactory cilia, fine hair-like processes that project up into the viscous mucous layer that covers the moist surface of the inner nose and greatly increase the membrane surface area available for odorant detection. Odorant receptors lie embedded in the surface of the cilia, ready to capture smells borne on the air you breathe.

Humans have around 350 distinct types of olfactory receptor proteins,[11] although each olfactory neurone carries only a single kind. But we can detect far more than 350 aromas: most people can distinguish many thousands of substances, often in tiny amounts. A good 'nose', such as an expert perfumier or sommelier, has even finer discrimination. Thus it is clear that there is not a specific receptor dedicated to a given odour. Rather, it is believed that each receptor recognizes a class of odour molecule (or a specific molecular feature), that a single odorant may bind to more than one receptor, and that it is the specific combination of receptors that are stimulated that enables us to discriminate smells. In the same way that the letters of the alphabet can be used to construct a vast vocabulary, so the different combinations of odorant receptors produce a cornucopia of pure odours. Scents are even more complex and varied as they are composed of many different odours.

It is widely believed that humans have a poor sense of smell. But tests show that we can detect some odours almost as well as dogs and much better than rats, and that we easily outperform highly sensitive measuring instruments. One reason for our supposed poor sense of smell is that we walk around with our noses high in the air, while scents are at their strongest close to the ground and quickly dissipated by air currents at higher levels, as can easily be seen by watching how a tracker dog follows a trail. Moreover, despite being able to recognize many different aromas, most of us are not very good at describing this difference in words. Yet the ability to identify a wine as distinct from all others is a highly complex and demanding task, and even those of us who are untrained find no difficulty in distinguishing the scents of oranges and lemons, which are simply mirror images of the same molecule, limonene.

When you smell a rose, the scent is wafted up to your olfactory epithelium, where the many different chemicals that make up the smell bind to their receptors on different sets of olfactory neurones. Precisely how odorants stimulate their receptors is still unclear, but it appears to be due to the different sizes and shapes of the odorant molecules. One idea is that they bind to the receptor in a lock-and-key fashion. Just as your right glove will only fit your right hand, so right-handed molecules will only bind to right-handed receptors; this explains why orange and lemon (which are left- and right-handed versions of limonene) smell different. Binding of an odorant to its receptor triggers a cascade of events in the neurone that leads to opening of a specific kind of ion channel – related to, but different from, those in the rods and cones – so giving rise to a current that in turn sets up a stream of action potentials in the olfactory neurone itself. These impulses pass along the olfactory nerve to a region of the brain known as the olfactory bulb, where they hand their signals on to other nerve cells in deeper regions of the brain. One of these is the limbic system, which is involved in emotion, which explains why smells can trigger such powerful emotions and memories.

As the olfactory nerve fibres run from the nose to the brain they pass through holes in the skull that form part of the cribiform plate. Consequently, a severe jar to the head may shear the nerves against the skull, severing or damaging the nerve processes, which usually results in a permanent loss of smell and, because smell and taste are intimately linked, it can also lead to a loss of taste.

Olfactory neurones that possess different kinds of receptors are randomly distributed across the olfactory epithelium. In the brain, however, they sort themselves out, with cells that express the same type of receptor all converging on the same place. The olfactory neurones are unique among nerve cells in that they turn over very rapidly. Each lives only about sixty days and is then replaced by a new neurone that differentiates from an olfactory stem cell. To preserve the map in the brain, replacement nerve cells that bear the same kind of receptors must always find their way to the same place in the olfactory epithelium. How this complex rewiring is achieved is still a mystery.

The King of Fruits

Like most sensations, if you are exposed to a constant smell you gradually become accustomed to it and eventually no longer perceive it. Most people fail to notice their own body odour, or even the perfume they are wearing after a while. But some smells linger longer than others. The durian is revered in South-East Asia as the most delicious of fruits. It is also one of the smelliest – so pungent, in fact, that it is banned from airplanes and hotels. I once came across a durian in a Chinese market in London and having heard of its reputation as the King of Fruits, I bought it and took it back with me on the train to Oxford. During the hour-long journey, the busy rush-hour carriage slowly emptied as the distinctive smell of the durian escaped from my bag. By the time I arrived, I was sitting in solitary state and the smell was overpowering – an indescribable mixture of smelly socks and rotting food that was so disgusting I could not tolerate the idea of having it in my house and instead left it at the lab. Next morning, when I entered the room I reeled back in shock, hit by an overpowering stench. The plan had been to taste the fruit at lunchtime, but long before that the smell had crept down the corridor and penetrated to the front lobby, and people were asking 'What's that funny smell?' Rapid action was needed. So, you may well ask, was it really so delicious? Alas, to me the taste was far less memorable than the stench and not particularly pleasant. I am not alone; the French naturalist Henri Mouhot remarked, 'On first tasting it I thought it like the flesh of some animal in a state of putrefaction.' Clearly, it is a taste that has to be acquired.

Touched

From the caress of a loved one, to the feel of the wind on our cheek or the crush of a bear-hug, touch plays an important part in all our lives. Sense organs in our skin respond to such mechanical forces with an electrical change, triggering nerve impulses that relay information back to the spinal cord and brain. Like other sensory nerves, impulse frequency is graded according to the stimulus strength, with lighter touches evoking fewer impulses than stronger pressures. Touch receptors also adapt to continuous stimulation, which explains why we do not notice the pressure of the clothes we wear.

Exactly how mechanical energy is transformed into electrical energy remains a puzzle, but it is clear that mechanically sensitive ion channels are somehow involved. Recent studies suggest that these channels are attached to the extracellular surface of the cell by a gating tether, in an arrangement similar to that found in the hair cells of the ear. Pressure on the cell membrane is thought to tug on the tether, distorting the structure of the channel so that it opens. The more the membrane is deformed, the more channels are likely to be activated and the greater the excitation of the nerve. Sometimes nerve endings sensitive to mechanical force are packaged into specialized structures that enhance their ability to detect changes in pressure or vibrations such as those that arise when you stroke your fingers across a rough surface. However, the end result is the same: a mechanical stimulus elicits an increase in action potential frequency in the sensory nerve.

Some Like it Hot

Our skin not only contains receptors sensitive to pressure, but also to temperature and painful stimuli. Bite into a habenero chilli pepper and it explodes in your mouth like a firebomb. Its burning pain comes from the chemical capsaicin it contains and different varieties of pepper have different amounts of capsaicin, which explains their very different potencies. In 1912, Wilbur Scoville calibrated the strength of chillies by measuring how much an extract of the pepper must be diluted until it was barely detectable when placed on the tip of the tongue. On the Scoville scale, the mild bell pepper notches up less than one heat unit, a jalapeño pepper has 2,500 to 5,000 units, and the famously incendiary Bhut Jolokia well over a million. The potency of pepper sprays used to repel grizzly bears, elephants and human attackers can be even higher: the weapons-grade pepper spray used by the Indian army cracks in at two million Scoville units.

When Mike Caterina and David Julius first isolated the capsaicin receptor, it turned out to be an ion channel. Binding of capsaicin opened the pore and stimulated electrical activity in the sensory nerve. The channel was also opened by noxious heat. So the reason chilli peppers taste so hot is that they open the same ion channel as high temperature, and because the brain cannot tell the difference between the two stimuli it interprets them both as heat. These channels are not just found in the tongue, they are also present in the skin of your fingertips, face and other sensitive parts of the body – as unfortunate men who have been chopping chilli very quickly find out if they forget to wash their hands before visiting the lavatory. Unlike humans, birds are not sensitive to chilli because they have a mutation in the channel that renders it less sensitive to capsaicin. This is highly advantageous to the plant, for their seeds are spread by wild birds. It is also why it is recommended to deliberately add chilli powder to bird food to deter squirrels from stealing it.

Just as chilli stimulates hot receptors, so other chemicals interact with receptors that sense cold, fooling the body into thinking the substance is cool. The minty, fresh taste of menthol, found in peppermint oil, arises from the fact it activates an ion channel that detects cold temperatures. This channel is structurally very similar to the capsaicin receptor and in fact we now know that there is a whole family of such channels, called TRP channels, each of which detects a different shade of temperature. Many of these channels are also sensitive to a range of pungent or painful chemicals – not just capsaicin, but substances such as wasabi (the hot Japanese horseradish), mustard, garlic and camphor.

Some snakes have exploited the ability of TRP channels to sense heat to produce natural thermal-imaging cameras that enable them to detect the body heat of their prey and so track their movements and strike accurately even in the dark. The Western diamond-backed rattlesnake, a pit viper, is unmatched in its sensitivity to infrared radiation, being able to detect a change in temperature as small as 0.01 °C. It has two exquisitely sensitive heat sensors, known as pit organs, which lie on either side of its head. These consist of spherical pits, open to the outside, within which a thin heat-sensitive membrane is suspended. Sensory nerve endings ramify through the membrane, their tips crowded with a type of TRP channel known as TRPA1, which serves as the heat sensor.[12] It is postulated that heat activates the TRPA1 channels, stimulating firing of the sensory nerve and alerting the snake that its prey – or a predator – is present. Vampire bats also use TRP channels to home in on their warm-blooded prey; they are found in specialized heat-sensing organs located around the bat's nose.

But TRP channels are not only used to sense temperature. Those sensitive to thermal extremes also serve as pain receptors and when they are stimulated we feel it hurts. This explains why it is difficult to discriminate between extreme heat and intense cold – between fire and ice. One feels only pain. As Shelley eloquently put it, 'the bright chains eat with their burning cold into my bones'.

Such a Pain

Pain can be extremely useful – it is a valuable alarm system that signals danger. It tells us that the pan is hot; that our toes are in the fire; that straining too hard will tear our muscles; that we have an infection or a wound. Without it you may be burnt, develop suppurating sores, or walk around with broken limbs, causing further damage. Pain also reminds us to allow damaged parts of the body to heal. A common side-effect of diabetes is the loss of sensation in the feet and legs. As a consequence, blisters, sores and minor injuries can go unnoticed, leading to infections that ultimately may necessitate amputation of the affected limb.

In addition to TRP channels, one of the ten kinds of human sodium channel is involved in the perception of pain. This channel, known as Nav1.7, fails to work in some people. Because their pain nerve fibres can no longer conduct action potentials, they are unable to feel pain, although their sensitivity to touch, temperature, pressure and so on is completely normal. This is far from a blessing, for pain is a valuable warning and bruised and broken limbs may go unnoticed in people who lack functional Nav1.7 channels. Indeed, scientists first identified the role of the channel in pain sensation by studying the family of a young Pakistani boy who made a living by stabbing knives into his arms, or walking on burning coals, in a gruesome form of street theatre. On his fourteenth birthday he jumped off a house roof to prove just how tough he was. He died from his injuries, which mercifully he was unable to feel.

Equally disabling is the opposite condition in which the Nav1.7 sodium channels are permanently activated. This condition is known as erythermalgia and it runs in families. Patients suffer episodes of intense debilitating pain, associated with redness and burning sensations in their hands, arms and legs. They complain that they feel as if hot lava had been poured over their body, as if their feet were on fire or of the sensation of walking on hot sand. These symptoms tend to be provoked by warm weather, exercise and the use of bed sheets, and many sufferers are unable to wear shoes because of the pain. It seems Nav1.7 sets the gain on pain – too much channel activity and you will be in permanent pain, too little and you will be always anaesthetized. Interestingly, a common variant in the Nav1.7 gene alters your pain threshold and could explain why the same stimulus feels more painful to some people than to others.

All pain comes from the brain. It is our brain that receives messages from nerve fibres, telling us we have stubbed our toe, and many brain areas are involved in our experience of pain; they tell us where the pain is, how much it hurts, and what kind of pain it is – sharp, burning or just a dull ache. Our perception of pain is also highly variable. Even if the input signal from our sensory nerve endings is identical, the way the signals are processed is powerfully influenced by our attention, mood and expectation and this can dramatically alter the pain we experience. Our emotions can make a placebo pill an effective painkiller even though it contains no active ingredient, and, conversely, fear of pain can sharpen its impact.

The main problem with pain is that once we have registered its message we are unable to switch it off. Worse still, in some unfortunate people the pain may remain even after the body has healed. Such chronic pain is very common, and is experienced by as many as 15 per cent of adults. This can be devastating and may ruin their lives. Billions of dollars are spent on pain medication every year, but many painkillers are not very effective and some of them, such as those derived from opium, have addictive properties. Better drugs are urgently needed, especially for treating chronic pain, which is often not ameliorated by current therapies. Because Nav1.7 is mainly confined to pain neurones, a drug that specifically blocks these channels might be able to tune out pain without causing side-effects.

What a Relief

When I was a child, I dreaded a visit to the dentist as it was often a painful experience. No longer. Modern dentistry has been completely transformed by the introduction of new and better local anaesthetics. Even the removal of a nerve from a root canal is painless – the worst sensation is the sharp pinprick as the injection is given, and that too is partially numbed by application of a topical anaesthetic. Most local anaesthetics act by blocking sodium channels, preventing conduction of nerve impulses from local nerve endings in the teeth to the brain. Dentists commonly use lidocaine because it acts very rapidly. The problem with such drugs, however, is that they do not just inhibit electrical activity in pain fibres, they also affect the other sensory and motor nerve fibres so that some hours after we have visited the dentist we still have a lopsided smile and our jaw feels numb. What is needed is an anaesthetic that is specific for the sensory nerves.

One way to find this is to identify the types of ion channels specific to sensory nerves and then find a drug that selectively blocks them. Currently, the best target seems to be Nav1.7, and several drug companies are currently seeking specific inhibitors of this channel. This is far from easy as the drug must also be able to penetrate the sheath surrounding the nerve, must not be broken down by the body too quickly, and preferably should retain its activity when taken by mouth. Developing any new drug also takes a long time and is extremely expensive. Consequently, it may be some time before we no longer suffer a frozen jaw after a visit to the dentist.

The Sensational Brain

Information from our sense organs travels via the sensory nerves to the brain encoded as electrical impulses. Bypassing the sense organs and stimulating the sensory nerves directly therefore evokes a sensation, as Isaac Newton vividly demonstrated in the mid-1660s. He records how he slid a bodkin (a small needle) between his eyeball and the back of the eye socket and found that when he pressed on it 'there appeared severall white darke & coloured circles'. It is not necessary to perform such a dangerous experiment, however, to see coloured circles – simply gently pressing on the closed eyelid will do it. The pressure stimulates the retina, and thus the optic nerve, and is seen as light. Direct electrical stimulation of the region of the brain concerned with vision has the same effect, even in the blind.

Newton also records how the 'circles were plainest when I continued to rub my eye [with the] point of [the] bodkine, but if I held my eye & [the] bodkin still, though I continued to presse my eye

[with] it yet [the] circles would grow faint & often disappeare untill I removed [them] by moving my eye or [the] bodkin'. As you will by now appreciate, a common theme in the nervous system is that the response to a continuous stimulus gradually weakens. We are pre-programmed to respond most strongly to changes in our environment and cease to pay attention if nothing new happens, a phenomenon that has a clear evolutionary advantage.

Sensory experience, then, is coded in electrical signals. It is the brain that interprets this barrage of nerve impulses, and deduces – on the basis of where they come from – what they mean. When the brain fails to attend to its inputs we may stare at the world but fail to see what is there, and illusions arise when signals conflict. Nor is the brain merely a receiver, for it can tune the sensitivity of our sense organs and modify the information they receive. Our perception of sights, sounds, scents and so on is thus the result of a two-way collaboration between our sense organs and the brain. So let us next look at the role the brain plays in this sensational dance and how it modifies and shapes the fractured information supplied by our sense organs, and weaves it together to produce a complete sensory picture of the world. To do so, we must first understand how the brain is wired up.

10

All Wired Up

Men ought to know that from the brain, and from the brain alone, arise our pleasures, joys, laughter and jests, as well as our sorrows, pains, griefs and tears. Through it in particular, we think, see, hear, distinguish the ugly from the beautiful, the bad from the good, the pleasant from the unpleasant.

Hippocrates, *On the Sacred Disease*

Hello. I am delighted to meet you – and particularly pleased you have made it this far. I hope it has been an interesting journey. Or perhaps you have just picked up this book and, riffling through the pages, have arrived at this one? Whichever it is, take a moment to consider how astonishing it is that I can communicate with you so easily across space and time. A vast number of obscure electrical miracles taking place in your brain enable me to do so.

As you read (or listen) to my words, the sensory cells in your eyes or ears are busily engaged in detecting information encoded as light or sound and transforming it into electrical signals. But that's only the start of the process – that information is then converted into a chemical signal and back again into an electrical one multiple times as it travels from sense organ to brain. And information that was first deconstructed into small manageable chunks of data is then processed and reassembled to form several sensory maps in the surface layers of your brain. Even more extraordinary, this information – this pattern of electrical signals flying around your nerve cells – is then interpreted as language and yet more sparks fly as you recognize my words and understand what I mean. If you like what I say, you might smile; and if you don't understand me or think my words facile you may by now be feeling frustrated or irritated – you may even (I hope not) be bored. And this too, these emotions my words trigger, are again produced by chemicals sloshing around your brain stimulating yet more nerve cells to fire. But the truly astonishing thing, most extraordinary of all, is that the person talking to you, writing these words – and indeed you yourself – is locked inside a small lump of jelly that fits neatly into your cupped hands and weighs no more than about 3 pounds: the brain. We are electrical beings, you and I, and we constitute no more than an unimaginably complex and continuously changing pattern of electrical and chemical signals.

The Little Grey Cells

Your brain is one of the most sophisticated machines on the planet. It has over 100 billion nerve cells and each of them communicates with many thousands of others. There are trillions of connections, as many as in the whole of the world's telephone system and far too many to fully comprehend. But the brain is not simply a great mass of interconnected nerve cells. It is a highly organized structure, with different parts being specialized for different purposes.

The most important bit of the brain – that responsible for our thoughts and actions – is the forebrain or cerebrum. It makes up about 80 per cent of the weight of the human brain and is divided into two mirror-image cerebral hemispheres, each of which primarily interacts with one side of the body. For unknown reasons, the wiring is crossed, with nerves from the left side of the body going to the right side of the brain and vice versa. The cerebral hemispheres are wired together by the corpus callosum, the brain's information super-highway: cut it, and you cannot name what lies in the left half of your visual field because the image is presented to the right side of your brain whereas language is processed on the left side of your brain.

The outer layer of the forebrain, known as the cerebral cortex, is made up of a thin sheet of nerve cells that is thrown into numerous folds to increase its surface area and enable more to be packed into the skull. Its highly convoluted structure makes it look rather like a walnut kernel. It is this four-millimetre thick layer of cells that mediates thinking, conscious actions, sensation, learning and memory, and different parts of it are specialized for different functions. Below the outer shell of nerve cells the forebrain is packed with nerve fibres that run to and fro wiring the nerve cells of the cortex together. So vast in number are these interconnections that the cortex spends most of its time talking to itself.

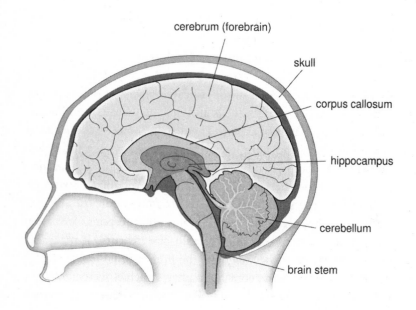

Cross-section of the human brain showing the major regions.

Below the forebrain lie regions of the brain that are involved in controlling the emotions, in regulating appetite and sleep, and that act as relay centres for processing information coming in from the sense organs and handing it on to the cerebral cortex. Even further down, at the base of the brain, sits the brainstem, which connects the upper parts of the brain to the spinal cord. It controls all your unconscious actions: here is where breathing, blood pressure, heart rate, digestion and so on are regulated. These regions may continue to survive and function even when higher brain functions have ceased, a condition known as a persistent vegetative state, in which the patient is often referred to as a vegetable. This bit of the brain is similar in structure to that found in many other creatures, and serves the same role: indeed, it is sometimes known as the reptilian brain.

Curled up at the back of the brain at the top of the brainstem is the cerebellum (or 'little brain'), which helps control balance and coordinates movements. It is involved in learning skilled motor tasks such as riding a bicycle, driving a car, and dancing *Les Sylphides*; damage it, and you cannot walk properly, but stagger around as if drunk.

Agatha Christie's famous detective Hercule Poirot was very proud of his 'little grey cells'. He was referring to the fact that although the living brain is a pinky-brown colour, when it is pickled the nerve cells turn grey, and hence are known as the grey matter. Nerve fibres, on the other hand, appear white and shiny when pickled (because of their myelin coats) and are called the white matter. But the brain does not consist solely of nerve cells. There are almost as many supporting cells, called glia, which help guide developing nerve cells on their way, supply them with nutrients, envelop them within a myelin sheath and guard them against infection. The delicate brain tissues are enveloped by membranes (the meninges) and shielded by a protective skull; within it, the brain floats in a sea of cerebrospinal fluid that cushions it and prevents it being damaged if the head is accidentally knocked, in the same way that the amniotic fluid protects the developing baby in the womb.

The brain has a large blood supply and many people will die, and an even larger number will be permanently handicapped, by blockage or rupture of the cerebral blood vessels, such as occurs in a stroke. Loss of the blood supply in this way leads to death of nerve cells in the local environment through lack of oxygen and nutrients and the build-up of toxic waste products. However, brain cells are not in direct contact with the bloodstream, but are protected by the blood-brain barrier. This is formed from the layer of cells that line the smallest blood vessels, which are so tightly knit together that they prevent substances leaking between the bloodstream and the cerebrospinal fluid. This blood–brain barrier is an important defence against noxious substances and infectious agents, such as bacteria and viruses, which drift around in the bloodstream.

The brainstem is directly connected to the spinal cord. When you decide to wriggle your fingers and toes the brain sends signals down the spinal cord and out along the peripheral nerves to command your muscles to move. The nerves that come out of the spinal cord in the small of the back and below serve the muscles of the legs; those higher up in the neck region signal to the arms. Damage to the nerves in the spinal cord means that electrical signals will be disrupted leading to paralysis and loss of sensation, as everything below the injury ceases to function properly. People who sever their spinal cord in the middle of their back can no longer walk but will still be able to breathe and move their arms. Break your neck, however, and you may be unable to move or feel anything in your arms as well. Depending on the exact location of the break, you may also be unable to breathe unaided.

Damaged nerve fibres in the brain and spinal cord never recover, leaving the patient permanently disabled. This was known even to the Ancient Egyptians, who declared that a person having 'a dislocation in a vertebra of his neck' is unaware of his legs and arms and cannot be treated. Over 3,700 years later, it is still the case. Not so for the peripheral nerves. My father severed the nerves in his fingers when adjusting the blade on an old lawn-mower and was left with no sensation in his fingertips, a devastating blow for a potter. Within a year or so, however, he was able to feel his fingertips once again as the nerves had grown back: but the regrowth was a slow process, inching forwards at less than two millimetres a day.

Seeing Single Cells

Individual brain cells are so tiny that it was not possible to see them until the invention of the microscope. Even then, the huge interconnected mass of cells within the brain and nerve trunks meant that special stains were needed to visualize individual cells clearly. In 1871, Camillo Golgi developed just such a stain.

While working as a medical officer in a psychiatric hospital in northern Italy, Golgi pursued his real passion – unravelling the anatomy of the brain – in a makeshift laboratory converted from an old kitchen. After a long series of attempts he discovered that a combination of potassium dichromate and silver nitrate stained a very few nerve cells randomly, but in their entirety. Paradoxically, the most important thing about Golgi's method was that it hardly ever worked, for the fact that only a few cells were stained meant that for the first time it was possible to see the spidery shape of a single nerve cell in all its glory, with its multiple, delicate dendrites and long, thread-like axon.

The great Spanish anatomist Santiago Ramón y Cajal subsequently made a series of stunningly beautiful drawings of nerve cells visualized using Golgi's silver-staining method. He was a gifted draughtsman and had originally wished to be an artist, but his father persuaded him to study medicine. In the event he combined both professions. Based on his observations, Cajal proposed that each nerve cell is a distinct entity and physically separate from its neighbour. This led to a dispute with Golgi, who had a different idea. In the end, however, Cajal turned out to be right.

While silver staining enables a small number of neurones to be visualized in exquisite detail, it is not possible to see how neurones are connected together. What is needed is some way of colour-coding adjacent cells with different stains. This was achieved in 2007, using genetic techniques to label neurones with multiple different colours. In the same way that a television uses only three colours to produce many different hues, three different genetically encoded

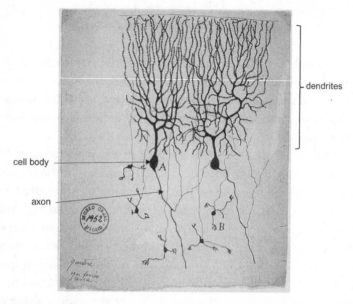

Drawing of Purkinje nerve cells (A) and granule nerve cell (B) from a pigeon brain by Santiago Ramón y Cajal, 1899. The cells were stained with the silver stain developed by Camillo Golgi. The small 'knots' on the dendrites are dendritic spines.

fluorescent dyes were used to paint the brain of a mouse. In one region of the 'brainbow' mouse brain over ninety different colours were distinguished, enabling the connections between neurones to be traced. It was not just clever science – it was also a work of art.

Taking the Brain Apart

Understanding how the brain is wired up, how information flows from one region to another, and how information is coded and processed is one of the most challenging and complex tasks in neuroscience. In an electronic circuit, such as that of a radio, the wiring diagram details the connections between individual components and how information flows around the circuit. There is only one animal on the planet for which the complete wiring diagram of the nervous system is known and that is a microscopic nematode worm called *Caenorhabditis elegans* that lives in the soil. It is a scientific supermodel and has received even more attention than the catwalk variety. Because it is so small and has such a simple nervous system, every single nerve cell and every connection is known. It has 302 neurones, about 5,000 chemical synapses, 600 electrical synapses and 2,000 nerve–muscle connections.

The enormous complexity of the human brain and the difficulty of identifying individual connections make construction of a similar circuit diagram for our own brains an almost insurmountable problem, never mind the fact that it would be different in each individual and would change as we learn new skills and have new experiences. Nevertheless, we are not entirely ignorant of how our brains work.

The idea that different bits of the brain are specialized for specific functions was first championed by Franz Joseph Gall in the early nineteenth century. By extensively examining the skulls of his friends, his patients and the inmates of local asylums and prisons, he came to the conclusion that different bits of the brain were associated with different mental attributes such as valour, cautiousness, ambition, wit and mechanical skill, and that this was reflected in the size and shape of the overlying skull. A charismatic speaker, he travelled throughout Europe delivering public lectures on his ideas, even giving a presentation to the German royal family. He also amassed a collection of 300 human skulls and over 100 plaster casts. But although phrenology – the practice of deducing a person's character from the bumps on their head – enjoyed a brief vogue, it has no basis in science.

The first real clues to understanding what jobs different brain regions carry out came from studying people with brain damage caused by injury or disease. One of the most celebrated of these individuals was Phineas Gage. On 13 September 1848, Gage was the foreman of a gang of construction workers building the bed of a new railway line outside the town of Cavendish in Vermont.

He was preparing to explode a large boulder and was using a long iron rod (one and a quarter inches in diameter, about four feet long and more than thirteen pounds in weight) to tamp down the blasting powder into a hole drilled in the rock. Alas, a spark caused by the iron striking the rock ignited the dynamite, which exploded, driving the rod straight through Gage's skull. It entered through the left cheekbone, damaging his eye, and exited though the top of his head, landing some twenty-five metres away, smeared with his blood and brains. Gage 'was thrown upon his back, and gave a few convulsive motions of the extremities' but rather remarkably he spoke within a few minutes, was able to sit upright in the cart that transported him to his hotel and then even walked up a long flight of stairs. The first doctor to examine him was disinclined to believe his story until Gage got up and vomited and 'the effort of vomiting pressed out about half a teacupful of the brain, which fell upon the floor'. A second physician, who arrived an hour and a half later, found Gage conscious and talking, but noted that both 'he and his bed were covered in gore'.

Although Gage recovered physically, it soon became clear the accident had changed him. Previously of well-balanced mind, friendly, energetic, hard-working, and a great favourite with his colleagues, he became obstinate, vacillating, uncooperative and indulged in 'the greatest profanity'. He was, said his friends, no longer the same man. What Gage's story shows is that our personality and emotions are linked to the function of certain brain regions. The damage to his prefrontal cortex had led to his inappropriate behaviour and loss of social inhibitions.

Another unfortunate whose ailment provided information about where different functions are located in the brain was Monsieur Leborgne, who was unable to say anything other than 'tan' when Paul Broca examined him in 1861. When Leborgne died shortly afterwards, a post-mortem revealed that a small region of his left cerebral hemisphere was damaged. Immortalized as Broca's area, this region of the brain is concerned with speech production. A few years later, Carl Wernicke discovered several patients with a different speech problem: although able to articulate words clearly and fluently, they simply spoke gibberish, a meaningless, incoherent rush of disconnected words, but with the syntax of the sentences more or less correct, such as in, 'I can't talk all of the things I do, and part of the part I can go alright, but I can't tell from the other people.' It is now recognized that this bit of the brain is involved in language comprehension. Wernicke's area lies some distance from Broca's area further towards the back of the brain.

For most purposes, the left and right sides of our brains are symmetrical. Language, however, is confined largely to the left side of the brain. A patient who has a stroke in their left cerebral hemisphere may therefore find themselves paralysed down the right side of their body and unable to speak. By contrast, a stroke on the right side of the brain can lead to paralysis of the left side of the body but usually has only a limited effect on speech. Fascinatingly, people whose Broca's area has been damaged by a stroke are often able to sing words that they cannot speak – it seems that singing involves an entirely different bit of the brain.

All Fired Up and Ready to Go

Another means of determining what function a specific brain region performs is to stimulate it directly with a small electric current. One of the first scientists to do so in a systematic fashion was Eduard Hitzig, who in the mid-1800s experimented on Prussian soldiers whose skulls had been shattered by bullets, leaving part of their brain exposed. Hitzig noticed that when a small current was applied directly to the brain it caused involuntary muscle contractions in the subject. Subsequent studies on dogs revealed that a small strip of cerebral cortex – now known as the motor cortex – controlled the movement of specific parts of the body.

In a similar fashion, sounds, sights and even the sense of touch are mapped onto the cerebral cortex. At the top of the brain sits the somatosensory system. Here, inputs from sense organs in your skin are neatly arranged so that all the signals from one location on your skin go to the same bit of the brain: legs, feet, fingers and toes each get their own areas. The most sensitive parts of the body, like the lips, fingers and genitals, are assigned larger brain areas, with more neurones, than less sensitive parts of your skin like the small of the back. Similarly, inputs from your eyes are mapped onto the visual cortex at the back of the brain, with the signals received by the same part of your visual field going to the same place, while sounds are organized according to frequency in the auditory cortex. Indeed, it now seems that there may be several such maps for each of the senses: like all good machines, the brain may have some built-in redundancy. The information is not wired straight through, however: it passes through many relay stations and is highly processed en route.

The ability to evoke sensation and action simply by stimulating a specific region of the brain is of considerable clinical significance. It is often used in brain operations to ensure, for example, that a surgeon removing a tumour removes the correct bit and nothing else. During this operation the patient is awake and able to say what they feel: it does not hurt as the brain has no pain receptors, and local anaesthetics are used to dull the pain fibres in the skin overlying the skull. Such operations can also yield useful information about where memories, words and information are stored.

Brain Waves

Early studies of the brain thus operated on much the same principle as a small boy with a new mechanical toy, who takes it to pieces to see how it works. More recently, non-invasive ways of looking at brain function have been devised, in which it is possible simply to watch what happens by recording brain activity.

The first of these techniques is the electroencephalogram (EEG), which is a record of your brain waves. Just as it is possible to record the electrical activity of your heart cells from electrodes attached to your chest, so it is possible to see what is going on inside your brain from multiple electrodes stuck to your scalp with conductive jelly. These pick up the minute voltage changes generated by the collective activity of millions of nerve cells in the surface layer of your brain. Your brain waves appear as oscillations in voltage that are constantly changing in frequency and amplitude as different regions of your brain surge into activity or fall silent. The EEG is much smaller and more difficult to record than the electrocardiogram and far more difficult to interpret. It's a little like trying to understand the complex relationships between people living in a large city by listening in on all their telephone calls simultaneously: the multiple unconnected conversations make little sense and the vast number of them means it is impossible to pick out individual conversations.

All this means that the EEG has rather limited value as a research tool. Nevertheless, it does provide a glimpse of what the brain is doing and it has been particularly useful in studies of sleep and epilepsy, both of which are associated with marked changes in the EEG. The first recording of a human EEG was made in 1924 by Hans Berger, but it was not until some years later that its clinical value became apparent when it was observed that an epileptic seizure appears as a dramatic increase in brain activity – an electrical earthquake, in effect. It was later found that the EEG can be used not only to detect a seizure, but also to provide an indication of where it originates.

The EEG is also used to monitor the depth of anaesthesia, and to distinguish between whether an individual is in a coma or is dead. In most countries, death is defined as the cessation of brain electrical activity and legally you are dead when your brain waves stop, even through the rest of your cells may survive for many minutes, or even hours, after brain death. Such a definition is not only sensible, but clearly also of importance for organ transplantation. It means that the heart of a deceased individual can be kept beating by life-support equipment, so that most of the organs remain alive and can be used to save another person's life.

Watching the Brain at Work

In the last few decades, new imaging techniques have transformed the way we can study the living brain. Brain-scanning methods can see deep within the brain and provide a far better picture of what is happening in different brain regions than the EEG. Unlike the EEG, however, they do not record the electrical activity of the brain directly. Instead, functional magnetic resonance imaging (fMRI) measures brain blood flow and positron emission tomography (PET) measures the metabolic activity of your brain cells. Both of these are believed to be related to the electrical activity of the brain, because the more active a nerve cell, the more energy it will consume and consequently its metabolism will increase to provide it. As nerve cells do not possess internal reserves of nutrients, the more active they are, the more glucose must be delivered via the bloodstream. Consequently, blood flow to that brain region is also enhanced.

fMRI has been invaluable for studying brain function, for it can be carried out on awake volunteers. It has revealed how the pattern of electrical activity in the brain changes during sleep, under anaesthesia, in migraine, epilepsy, and a multiplicity of everyday tasks such as learning, memory, speech – even thinking. Simply by scanning someone's brain while they are asked questions, shown pictures, or asked to think about something, we can identify which bit of the brain is involved. Ask a person to imagine playing tennis and the blood supply to their motor cortex increases, as they think of smashing a high lob or a sizzling serve. Broca's and Wernicke's areas light up when you speak, confirming what was found from studying brain-damaged patients, and the reward centres of the brain spring into action when a smoker thinks of a cigarette.

Brain-scanning technology is transforming our understanding of how the brain works and what we think about ourselves. But it is worth remembering that the smallest brain region that can be distinguished in such scans still contains many hundreds or thousands of neurones, and what is detected (indirectly) is their summed activity. Thus there remains a huge gap between our highly detailed knowledge of what happens at the level of a single nerve cell, and how individual nerve cells are wired together to produce the electrical activity of our brains.

MRI and PET scanners are also invaluable clinical tools. Damaged areas of the brain can be easily identified and tumours or regions prone to epileptic seizures can be located. If an operation is needed, having a detailed picture of the precise location of the problem and its relation to crucial regions of the brain means it is far easier to avoid collateral brain damage.

Recently, a team of scientists of Cambridge and Liege universities have shown that it is possible to communicate with people's brains directly, simply by asking them to answer 'yes' or 'no' to a question and then looking at their brain scans. Not that it is possible to determine if someone is simply thinking 'yes' or 'no', but if you are asked to envisage playing a game of tennis if your answer is affirmative, it is possible to detect a response in your motor cortex, and if you are asked to think of navigating around your house if your answer is negative, then a different region of your brain lights up. The patterns of brain activity are so distinctive that even an untrained observer can identify the subject's response with almost 100 per cent accuracy. While it seems somewhat uncanny to be able to talk to someone this way, even more scary is the fact that four out of twenty-three patients who were believed to be in a persistent vegetative state were also able to give correct answers to questions, suggesting they may be at least minimally conscious and able to hear, but are totally cut off from the world because they cannot move at all, not even flicker an eyelid.

How the Brain Sees

All these studies serve to show that different regions of the brain are specialized for specific functions. The big mystery is how information is encoded and processed by the brain and how the different bits of the brain communicate. While this is far from understood, dramatic progress has been made in the last fifty years. Consider a single example – that of vision.

Sight is a collaboration between our eyes and brain, for sensory experience does not arise from our sense organs alone. Open your eyes and you see a three-dimensional technicoloured world, but what the retina actually detects is a colourless, distorted, upside-down image that it translates into a myriad of electrical signals. Some processing of these signals takes place in the eye, and multiple messages are then transmitted via the optic nerve to the brain, where different regions act as relay and processing stations. Ulti-mately, the information ends up as electrical impulses in the neurones of your visual cortex, which sits at the back of your brain.

Here the electrical signals are pieced together to create images and meaning. Nerve cells that respond to the same type of visual signal are localized to the same area of the visual cortex. Specific neurones are assigned different tasks. Some nerve cells seem to be specialized for detecting movement, others fire only when a human face is detected and some, known as mirror neurones, fire both when an animal acts and when it observes the same action per-formed by another. Once the image is recognized, signals are sent to the amygdala, the emotional core of the brain, where its signifi-cance is evaluated. Is this a lover about to embrace you, or are you in fact about to be mugged? Is this the bus you have been waiting for? Or are you just looking at a beautiful landscape?

You must then decide if what you see requires action. This involves signals being sent to the prefrontal cortex, the executive region of the brain, where you decide, for example, whether it is worth sticking out your hand to signal the bus to stop. If so, then more signals pass to the motor cortex, which instructs the necessary muscles to move your arm. Those signals that came in from the eyes via the optic nerve thus result in a multitude of complex mes-sages that whizz back and forth around the brain. Bear in mind that we have not yet even considered how such visual information is integrated with that coming from other senses to build up a complete sensory picture of the world, or how that picture may be laid down as memory.

Our vision, of course, cannot be trusted. We do not always see what we think we do, as many optical illusions testify and many artists have exploited. Seeing is not believing because of the way our brains process information. We constantly make predictions about the world – anticipating, for example, where a squash ball will bounce, so that we can move to meet it before it lands. Illusions happen when the models we construct inside our heads, against which we unconsciously measure sensory information, do not match reality. In the left panel below, we see a non-existent inverted triangle, as our brain unconsciously fills in the missing lines. In the central panel, perspective tells us that the railway lines are receding into the distance and thus we see the light bars as a different size – even when we are told they are identical. The right-hand panel is seen either as two faces or as a candlestick, but never both; clearly, the image does not change – it is the way in which the brain interprets it. As such illusions reveal, our perception of reality is a construct of both the brain and the sense organs.

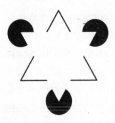  

This is particularly easy to demonstrate in the case of colour vision. White paper looks white even in yellow light because we are used to it being white. Further, as Patrick Heron's wonderful paintings show us, we judge colours by the company they keep; the same yellow, for example, looks different to us when it sits next to different colours. Early painters exploited this phenomenon to create the illusion of a colour for which there was no pigment then available (such as mauve). We can even perceive colour that is not there: a black-and-white image appears in colour when rotated rapidly. Conversely, if blood flow to the visual cortex is restricted the world turns grey – as can happen to boxers with head injuries.

Similarly, blindness does not always result from injury to the eyes. It can also be caused by damage to the visual processing regions of the brain; for example, by a stroke. Remarkably, some people who can see nothing and believe they are blind are able to 'guess' correctly what is sitting on a table in front of them, or pick up the right object when asked to do so. Such 'blindsight' demonstrates that we can see things without being consciously aware of doing so. Thus there appear to be at least two pathways by which visual information is processed in the brain, one that is linked to conscious awareness and one that is not.

Pay Attention Now!

The brain constantly filters the information it receives. Consider. Only the centre of our visual field is actually in focus, yet we see the whole of it in sharp definition. This happens because our eyes are constantly moving, focusing on different parts of the visual field, and the brain pieces the bits together into a coherent picture. We are blissfully unaware of what is happening because the brain ignores any visual inputs during the time our eyes are moving. This explains why if you stare in a mirror you will not see that your own eyes are constantly flicking from side to side – but another person will do so. Likewise, we 'tune out' background conversations and hear only the person we are talking to – unless we hear our name spoken, at which point our attention suddenly shifts. Our ability to attend to the most important information and disregard the irrelevant is very valuable, but it can also fool us.

I vividly recall one evening when I and a bunch of other scientists were asked to watch a film of a ball game between two teams, one dressed in blue and one in red. This film is now familiar to many people, but at the time it was new to me. We were instructed to count the number of times each team touched the ball. I was mortified when at the end of the film the lecturer said the number was of no consequence, what he really wanted to know was how many of us had seen the gorilla. Gorilla?! I had seen nothing, but to my surprise four people claimed to have done so, and when the film was replayed it was obvious – a man dressed in a gorilla suit walked into the centre of the screen, thumped his chest several times, and then strode off. How could I have missed it? It was an impressive demonstration that by focusing my attention elsewhere, my brain had ignored other information it had received.

The Gift of Coloured Hearing

Imagine seeing sound and hearing colours; a common experience when taking certain psychedelic drugs perhaps, but some individuals have this gift without recourse to pharmacology. One of the most famous was physicist Richard Feynman, who wrote, 'When I see equations, I see the letters in colours – I don't know why. As I'm talking, I see vague pictures of Bessel functions from Jahnke and Emde's book, with light-tan j's, slightly violet-bluish n's, and dark brown x's flying around. And I wonder what the hell it must look like to the students.' Another synaesthete with the gift of 'coloured hearing' was Vladimir Nabokov, who painted a vivid picture of the colours of the alphabet, with the green letters including the 'alderleaf f, the unripe apple of p, and pistachio t' and the blue group the 'steely x, thundercloud z, and huckleberry k'. As is clear from these descriptions, the synaesthetic experience is specific to the individual and words and letters do not always take the same colour – x, for example, was seen as blue by Nabokov and dark brown by Feynman.

Nor is the phenomenon confined to coloured letters. The great jazz musician Duke Ellington fused timbre and colour, writing, 'I hear a note by one of the fellows in the band and it's one color. I hear the same note played by someone else and it's a different color. When I hear sustained musical tones, I see just about the same colors that you do, but I see them in textures. If Harry Carney is playing, D is dark blue burlap. If Johnny Hodges is playing, G becomes light blue satin.' And Liszt astonished the Weimar orchestra when he requested, 'O please, gentlemen, a little bluer, if you please! This tone type requires it!' Other individuals may taste words or musical keys. They have an extra dimension to experience which most of us lack.

In synaesthesia, the palette of the senses becomes mixed. This melding of the senses does not occur in the sense organs, but in the brain, although quite how and where it is achieved is unclear. Brain scans have shown that the region of the brain concerned with processing colour information (the fusiform gyrus) sits right next door to that involved in number processing, suggesting that cross-wiring may be the reason why Nabokov, Feynman and others like them see coloured numbers. Presumably something similar happens when other senses get so gloriously muddled.

Synaesthetes may not be aware of their unusual abilities for some time, as they do not expect that others see the world a different way. Only when they discover, for example, that their friends do not remember phone numbers by their colour do they find out. In Nabokov's case it came up at the age of seven when he was building a tower of alphabet blocks and casually remarked to his mother that the colours of the letters were all wrong. They then discovered that she too saw tinted letters; synaesthesia, it turns out, often runs in families.

Migraine

I am no synaesthete, alas, but I too occasionally have unusual visual experiences. In my case, it generally starts with the world becoming blotchy, rather as if I were looking through a car window during a rain shower when the windscreen wipers are not working. At other times, I see brightly coloured stars or shimmering zigzags that march across my visual field screaming 'Wow! Zap!' like something out of a comic book or a Roy Lichtenstein painting. Much as I may enjoy this spectacular lightshow I fear its sequel, for it heralds the onset of a migraine. Soon I am feeling extremely nauseous, usually vomit, become acutely sensitive to light and then follows the accursed one-sided headache – so severe I feel unbelievably ill. The only thing to do is to hide away in a darkened room and wait it out.

I am not alone. Many others struggle with this most malign of headaches although not all experience the preceding aura. The remarkable colours and distorted visual perceptions have been described by, perhaps inspired, many writers and painters, among them Virginia Woolf and Lewis Carroll. But as Virginia Woolf once said, 'English, which can express the thoughts of Hamlet and the tragedy of Lear, has no words for the shiver and the headache.' It is an unaccountably nasty experience. Hildegard von Bingen's paintings and descriptions of visions of intense points of light and extinguished stars suggest that she too was a migraine sufferer.

One explanation of the strange visual phenomena associated with a migraine aura is that they occur because the electrical activity of the visual cortex is stimulated and that as this wave of excitation spreads across the cortex it generates illusory colours and perceptions. But this idea, like the origin of the headache itself, remains controversial. What is clear is that some unlucky families have severe forms of migraine that are caused by mutations in ion channels, including both calcium and potassium channels, that lead to enhanced electrical activity. In some people this activity is so intense that it eventually damages the nerve cells themselves so that ultimately they may even be unable to walk. There is evidence that enhanced electrical activity is also at the root of the problem in more common types of migraine. It is far from pleasant when suffering from a horrendous headache to reflect that it may also be damaging your brain.

The Balance of Power

Modern science, as we have seen, can now provide us with a crude map of the brain. In broad terms we know which regions are involved in processing different types of information. We can eavesdrop on the living brain while it is performing various functions, and see which bits are activated or suppressed. But what happens at the levels of the nerve cells themselves? How are they wired together and how do they talk to one another? The crucial element in the giant jigsaw that is the brain are the synapses, where, in Cajal's words, nerve cells exchange 'protoplasmic kisses, the intercellular articulations, which appear to constitute the final ecstasy of an epic love story'.

For synapses are not confined to the nerve–muscle junction. They are also found between nerve cells and gland cells and, very importantly, between nerve cells and nerve cells. There are several hundred trillion synapses in the brain, and millions more in the spinal cord. Typically each brain neurone will make contact with several thousand other cells. It is this intricate tapestry of connections that enables the complex behaviours of higher animals, including you and me.

Some of these synapses are excitatory, with the action of the transmitter being to excite the next cell, stimulating it to fire an impulse. Others, however, are inhibitory and the transmitter they release turns off the next cell in the chain, damping down its activity. Most nerve cells receive multiple excitatory and inhibitory inputs and their output reflects a balance between the strengths of these competing signals. In such a system, timing is crucial. An inhibitory signal will be ineffective if it arrives after an excitatory one, and an excitatory signal may find its passage blocked if it arrives at the same time as an inhibitory one. Further complexity is added by the fact that synapses present on the pre-synaptic nerve terminals may prevent the release of a transmitter. Therefore it is no easy matter to predict the response of even a single cell in an electric circuit.

In contrast to the nerve–muscle junction, where acetylcholine is the main transmitter, a cornucopia of different transmitters and their receptors are found in the brain. The main excitatory transmitter in the brain is glutamate – better known perhaps as the artificial flavouring in many Chinese dishes. Too much glutamate leads to over-stimulation of the target cell and can cause its death. Glutamate is therefore something of a Jekyll-and-Hyde molecule, essential for normal brain activity, but with the potential to destroy nerve cells completely. Consequently, cells have evolved ways to rapidly reduce the brain glutamate concentration to a low level, and transporters that capture extracellular glutamate and pump it back into cells are found at all glutamate release sites. The main inhibitory transmitter in the brain is gamma-amino-butyric acid (GABA), and that in the spinal cord is glycine. Many problems arise if any of this triumvirate of transmitters, or their receptors, are altered, or if we consume drugs or toxins that interfere with their function.

On the Horns of a Dilemma

In 1974, and again in 1997, Ethiopia was gripped by famine. The Western world watched in shock as horrific images of emaciated children and adults were beamed straight into their living rooms by satellite television. Food aid programmes burgeoned.

Unknown to these observers, a second tragedy was also unfolding. Many of the sufferers were reduced to crawling, not through weakness induced by starvation, but as a result of a poison contained in the only food available to them. A local doctor, Haileyesus Getahun, travelling in the remote and beautiful highlands of northern Ethiopia, described how he met a family of six, who 'attempted to offer the traditional respect by bowing but unfortunately to no avail. All but one girl were crawlers'. The mother had to be tied with a rope to the wall of the hut to prevent her from falling down while grinding grain, and the family was totally dependent on relatives and neighbours for survival. They were tragic victims of a recent epidemic of lathryism – a crippling paralysis caused by eating the grass pea *Lathyrus sativa*.

The grass pea has been cultivated in South Asia and Ethiopia for over 2,500 years. It is a popular crop because it is cheap and easy to grow, withstands drought and flooding, is resistant to insect attack and produces a good yield of highly palatable seeds. It is also often the only food plant to survive extreme drought. Thus it sounds like the perfect crop for areas prone to famine conditions – except for one small catch. The plant contains high levels of a potent nerve toxin with the unpronounceable name of β-*N*-oxalylamino-L-alanine (usually abbreviated BOAA). BOAA specifically damages the motor nerves that control movement for, like glutamate, it is an 'excitotoxin' – a poison that acts by stimulating nerve cells so much that they die.

BOAA binds to glutamate receptors on motor nerve cells in the brain. These receptors are themselves ion channels, and glutamate binding opens the pore, allowing calcium ions to flood into the cell. Because too much calcium is toxic to nerve cells, the continuous excitation of glutamate receptors by BOAA eventually results in nerve cell death. As a consequence, those who eat the grass pea for long periods develop a flaccid paralysis of the legs and are reduced to crawling. There is little or no recovery, even after consumption ceases.

Lathryism is the oldest neurological disease known to man. As long ago as 400 BC, the famous Indian physician Charak recognized that it was associated with excessive consumption of the grass pea, and about a century later Hippocrates wrote that at Ainos 'all men and women who ate peas continuously became impotent in their legs'. The first account to clearly establish the link between lathryism and grass-pea consumption was that published in 1844 by Major-General William Sleeman in his book *Rambles and Recollections of an Indian Official*, which described an outbreak of lathryism affecting both cattle and humans in the Saugor district of Central India.

Despite this knowledge, however, tragedies continued to happen. During World War II, the inmates of a German concentration camp at Vapniarca on the Ukrainian border were given a daily ration of grass pea and bread. Within three months over 60 per cent of them had developed lathryism. The cause was eventually identified by one of the prisoners, who was himself a victim, and the problem was solved by removing the grass pea from the prisoners' food. This incident led to recognition that consumption of the grass pea as a major part of the diet over a three-to-four-month period is required to cause paralysis.

Unfortunately, in some circumstances, people are faced with a stark choice between starvation and lathryism. In 1997, the grass pea was the only crop that survived the severe drought in Ethiopia, and it was consumed in many different forms. Although the plant was known to be harmful, the exact nature of the problem and how to prevent its effects were only poorly understood. Health workers were at a loss and advised the community to avoid contact with the steam or the drained-off water of the cooked grass pea – a common misconception that is inaccurate and of little value. No information pamphlets were available, and many communities were so remote they could only be reached by mule or on foot. The epidemic continued for a further two years, until grass-pea consumption fell.

Too Much of a Good Thing

Lathryism is not the only disease caused by hyper-stimulation of glutamate receptors. Early one morning in the summer of 1961, the coastal town of Capitola in California was attacked by an enormous flock of sooty shearwaters. Hundreds of them besieged the town, slamming into houses, dive-bombing people, falling lifeless from the sky and staggering around vomiting fish. People woke to find the streets littered with dead seabirds. The event so intrigued Alfred Hitchcock that it influenced his production of Daphne du Maurier's short story, 'The Birds', and he even included a reference to the Capitola attack in his film. Investigation of similar incidents that occurred subsequently showed that they were due to domoic acid, a poison that is made by phytoplankton and accumulated at high concentrations by creatures further up the food chain. The shearwaters had picked up the poison from the anchovies on which they had been feeding.

Shellfish also accumulate domoic acid. This led to a mass outbreak of shellfish poisoning in Canada in 1987. More than 200 people who had eaten blue mussels were affected. In addition to vomiting and diarrhoea, many suffered disorientation, memory loss, seizures and coma and about a quarter of them also lost their short-term memory, in some cases permanently. Autopsies on the brains of four people who had died showed that their hippocampal neurones (which are important for memory) had been destroyed. Domoic acid kills neurones by binding very tightly to glutamate receptors of the kainate variety and causing them to open; the result is an influx of calcium ions that kills the nerve cells.

A number of fungi also produce compounds that activate glutamate receptors, which explains why their consumption causes dizziness, delirium and euphoria. The sodium salt of glutamate (monosodium glutamate or MSG) is a flavour enhancer that is added to numerous foods, often under the name of 'hydrolyzed vegetable protein', to improve palatability. It has received a bad press ever since it was suggested to provoke nausea, dizziness and a splitting headache – the famous 'Chinese restaurant syndrome'. Most of the brain is protected from the effects of MSG by the blood–brain barrier, but a few nerve cells lie outside this protective barrier. In young mice, these cells are vulnerable to the toxic effects of high levels of MSG and if they eat too much they become extremely obese as the neurones that regulate body weight are selectively killed. Numerous safety studies, however, have shown that MSG has no adverse effect in humans at concentrations very substantially higher than those used as a food additive, and double-blind trials even failed to reproducibly demonstrate that MSG causes Chinese restaurant syndrome. Nevertheless, this did not stop MSG being the focus of the 'Soup Wars' campaign. The battle began when the Campbell Soup Company ran an advertisement featuring its own

soup as being made of 100 per cent natural ingredients and unflat-
teringly contended that its competitor's Progresso line of soups
tasted of MSG. The General Mills Company, who make Progresso
soups, counter-attacked, pointing out that in fact many of Campbell's soups contained MSG, whereas many of their own soups did
not. It also declared it was removing MSG from all its soups and
challenged Campbell to do the same. And so it went on.

Scared Stiff

Sudden noises can make all of us jump. But imagine if every time
you were startled your muscles seized up so severely that you froze
rigid and toppled off your chair, or fell flat as a plank, like a clown in
a circus act. This can happen to people with startle disease. Because
they find their arms become stiffly clamped by their sides, they are
unable to protect themselves when they fall and consequently may
suffer multiple injuries. Babies with startle disease can be so severely
affected that their spine arches backwards, as if in a tantrum, and
their respiratory muscles seize up so that they suffocate and die.
The disorder is sometimes known as stiff-baby syndrome.

The symptoms of this extraordinary disease are similar to those
of strychnine poisoning, which provides a clue to its origin. The
common cause is loss of glycine receptor function – either due to a
mutation, as in startle disease, or to inhibition by strychnine. Glycine
is one of the main transmitters at inhibitory synapses in the spinal
cord and brain stem. It is released from inhibitory nerve cells and
interacts with glycine receptors in the post-synaptic nerve cell membrane, opening an intrinsic ion channel that is permeable to chloride.
This damps down the electrical activity of the target cell and prevents it from responding to excitatory inputs. Such inhibition is
essential for normal function. The muscles that move our limbs tend
to come in opposing pairs, one of which flexes the limb and the
other which extends it. It is crucial that when one muscle is stimulated to contract the other relaxes, for if both contract simultaneously

the limb becomes stiff and cannot move. People with startle disease cannot respond to the glycine released from their inhibitory nerves, so their opposing muscles fail to relax. As both muscles contract simultaneously, they become rigid when surprised.

Although this disease has some similarities with that of the Tennessee myotonic goats described earlier (in both cases the muscles stiffen up) it has a very different origin. Startle disease is a problem of the central nervous system, which fails to provide the muscles with the correct signals. The muscles themselves are normal. In contrast, the nerves of the myotonic goats are unaffected; it is the muscles that are defective.

'The Mysterious Affair at Styles'

Late one night, Mrs Emily Cavendish, a wealthy widow, was found dying at her Essex country manor house, Styles Court, from what later proves to be strychnine poisoning. Agatha Christie's famous detective Hercule Poirot unravels a series of intricate twists and turns in the novel to prove that her new husband and his lover are the culprits. Strychnine has been used in many famous cases of poisoning, both real and fictional. The Lambeth serial killer, Dr Thomas Neill Cream, invited prostitutes to take a drink with him, spiked their drinks with strychnine and left them to die in agony. As strychnine is one of the most bitter substances known, the drinks must have been sweetened to disguise its taste, or the girls must have been sufficiently inebriated not to have detected it. Strychnine was also once used as a rat poison.

Strychnine poisoning resembles startle disease because the drug blocks glycine receptors, rendering them non-functional. The toxin was first isolated from the beans of the plant *Strychnos ignatia*, which was named after Saint Ignatius of Loyola, the founder of the Jesuits. It is also found in the seeds of the Strychnine tree (*Strychnos nux-vomica*). Intriguingly, strychnine was once used as a stimulant, albeit at concentrations lower than that which cause severe toxicity. As might be expected, this sometimes led to accidental overdoses. A medical student, writing in 1896, described how he took it while studying for an examination because he was feeling run down. His calf muscles began to stiffen and jerk, his toes curled up, he saw flashing lights and he broke out in a cold sweat. He wrote that he 'knew something serious was developing' and so crawled to his medical case and drank potassium chlorate (an anaesthetic). He quickly lost consciousness and fell into a profound sleep 'awaking in the morning with no unpleasant symptoms' but a desire to be on the move and a temporary stiffness in the jaw. Not, one imagines, an experience he wished to repeat.

Brain Storms

Loss of inhibition in certain brain circuits can also trigger epilepsy, a sudden uncoordinated burst of excess electrical activity that resembles an electrical storm of the brain. Fyodor Dostoyevsky was probably the most famous epileptic in history. He recorded 102 seizures in his notebook and incorporated his experiences into his novels. Fits or seizures tend to be unique to the individual, for there are many types of epilepsy and many causes, but they can be broadly grouped into two main kinds. In petit mal or absence seizures, the patient goes blank for a few seconds, staring into space and seemingly switching off from the world around them. More dramatic are the convulsive seizures in which the victim's limbs twitch and shake uncontrollably because the electrical storm influences the nerve cells that regulate their limb muscles. In some people the convulsions are highly localized and affect only a small group of muscles, whereas others may experience a grand mal seizure that is associated with generalized convulsions and often loss of consciousness.

Epilepsy has been known since antiquity. Hippocrates referred to it as the 'sacred disease', and correctly argued that it was caused by a disturbance of brain function. Nevertheless, for many centuries the view that epilepsy was a medical problem coexisted with the idea that epileptic individuals had been cursed by the gods or were possessed by evil spirits. Epileptics were often ostracized and by the sixteenth century were even branded as witches. Gradually, it was recognized that epilepsy was an illness, but it still carried a negative aura. When Prince John, the youngest son of King George V, developed epilepsy he was hidden away in a cottage on the Sandringham estate. Fortunately, these days there is no stigma attached to the disease.

The origins of epilepsy are still not fully understood. In some cases it results from a traumatic brain injury, a tumour that presses on the brain, or brain damage sustained during birth. In other cases it is inherited and caused by mutations in specific genes, many of which are ion channels. Most of these mutations impair the electrical activity of inhibitory nerve cells that normally exert a brake on brain activity. Release the brake and the brain goes into overdrive as excitatory circuits then become over-stimulated.

Early treatments for epilepsy were bizarre, ranging from Pliny's advice to drink the blood of gladiators to Robert Boyle's suggestion to eat crushed mistletoe, 'as much as can be held on a sixpence coin', when the moon was full. A landmark in therapy came when it was recognized in the late nineteenth century that removal of the trigger area could help treat epilepsy. Surgery is not always possible, however, and other parts of the brain may be damaged when removing epileptic foci. Current therapies often involve drugs that reduce the frequency and intensity of seizures. Many act by enhancing the release or action of the inhibitory transmitter GABA, which prevents excess electrical activity by holding nerve cells at a more negative level. Others suppress the activity of excitatory neurones directly by acting on the sodium and potassium channels involved. However, as epileptic seizures can damage the brain, such therapies may only have partial success unless the patient can be treated early.

Some unfortunate children have intractable epilepsy that is unresponsive to drug therapy and involves parts of the brain inaccessible to surgery. An old treatment that is surprisingly effective in some of these patients is to severely restrict their consumption of carbohydrates. Known as the ketogenic diet because it leads to the rise of metabolic by-products known as ketone bodies in the blood, it stops most seizures in about a third of patients and reduces their frequency in a further third. Why it works is far from clear, but the patients and their parents are not bothered about that. It is not an easy diet to stick to, however, and consumption of a single chocolate bar or other carbohydrate treat can precipitate a seizure.

Wiring the Brain

As this chapter has shown, the way the brain is wired up determines the intricate pattern of electrical impulses and 'chemical kisses' between cells, and so influences how we move our limbs and sense our environment. But it has a still more important and extraordinary function. As we shall see next, it determines our emotions, thoughts, personality, consciousness – our very sense of self.

11

Mind Matters

O body swayed to music, o brightening glance,
How can we know the dancer from the dance?

W. B .Yeats, 'Among Schoolchildren'

Joy, sadness, fear, anger, exhilaration, despair; our emotions fluctuate like sunshine and clouds in a British summer. They influence
our thoughts, dictate our actions and form the basis of our personalities. But we are not mere puppets of our emotions. We are also
capable of reasoned argument, of rational thought and action, of
creative ideas that seem to come from out of the blue. Contrary to
the mediaeval view, there is not some homunculus sitting in our
brain pulling the strings. Rather, blind evolutionary forces have
shaped our brains so that everything we think, feel and do is gov-
erned by the electrical and chemical events taking place in our nerve
cells. It may seem uncomfortable to consider that your thoughts
and feelings are determined simply by clouds of chemicals washing
through your brain, and by the changing patterns of electrical activity they produce. Yet with a moment's thought you will recognize
that this is indeed the case, for drugs, hormones and diseases that
alter the levels of neurotransmitters in our brain affect us deeply,
transforming our emotions and our behaviour.

A small amount of alcohol, for example, may usher in a more
outgoing personality, cause us to behave irrationally, or sink into
melancholy. Women's moods may fluctuate with their menstrual
cycle. Regular running can produce a high so pleasurable that aficionados become stressed and irritable if they are prevented from
exercising. Adenosine, administered to control the heart rate, has
the extraordinary side-effect of producing a transitory feeling of

impending doom so severe that the patient may feel they are about to die. Parkinson's disease is noteworthy for its high incidence of associated depression. Syphilis causes marked changes in temperament, most famously in King Henry VIII. Simply stimulating certain regions of the brain can produce euphoria, anger – even, it has been claimed, spiritual experiences. All human emotions have their origins in the electrochemistry of the brain, and an intricate tapestry of chemical and electrical signals governs our every thought and action.

This penultimate chapter considers how neurotransmitters influence our moods, our memories and our thoughts, and how drugs of abuse enhance or mimic their effects. It looks at how our personalities are shaped by the electrical activity of our brain and considers what happens to us during sleep and anaesthesia. And it addresses the question that has perplexed mankind for centuries – what is consciousness and who, exactly, am I?

What a Pleasure

We are pre-programmed to seek pleasure. Food, sex, drink, exercise – all produce feelings of enjoyment that drive us to seek more. But our impulse to do so is more than hedonism or sheer sensual delight; it is a way of ensuring that our species survives. All pleasurable experiences stimulate the reward centre of the brain. This consists of several distinct brain regions, including the nucleus accumbens, the amygdala and the ventral tegmental area, which are wired together by a group of nerve cells known as the median forebrain bundle. Dopamine, one of the most crucial neurotransmitters in the brain, is intimately involved in desire and addiction. Pleasurable experiences such as sex, love and food trigger the release of dopamine in the brain's reward centre, which increases nerve cell electrical activity, reinforcing our sensation of pleasure and coercing us to have yet another chocolate or glass of wine – too much in some cases. Many drugs of addiction act by increasing the concentration of dopamine in the nucleus accumbens, thereby producing feelings of euphoria.

Long ago when I was just a teenager I went to see a film at the local cinema with a school friend and her family. The queue was enormous and it was clear that we were unlikely to get in. 'Never mind', said my friend's mother, 'let's go home and try the cocaine.' This was not as outrageous a suggestion as you might imagine. Her son had just returned from South America with a bag of coca leaves. These have been chewed by Peruvian Indians for over 8,000 years, mainly because alkaloids in the leaves act as an appetite suppressant and help keep them awake. I, however, did not find it a stimulating experience – all that happened was that my lips and tongue were slowly and mildly anaesthetized, rather as if I had been to the dentist. Nothing else. Perhaps this may have been because I was far too nervous to take more than the tiniest bite of a leaf or two: even then, cocaine, which originates from coca leaves, had a fearsome reputation as a drug of abuse.

Initially, however, it was widely lauded. Sigmund Freud regularly took cocaine while writing *The Interpretation of Dreams*, as he found that cocaine caused 'exhilaration and lasting euphoria' and had such a 'wonderful stimulating effect' that 'long-lasting intensive mental or physical work can be performed without fatigue'. In the nineteenth century a cocaine-laced drink, Vin Mariani, hailed as a tonic for body and brain, was such a favourite of Pope Leo XIII that he awarded it a special gold medal and appeared on a poster extolling its virtues. A pinch of coca leaves was also added to the original brew of Coca-Cola, along with extracts of the kola plant (hence its name). The power of cocaine to banish tiredness was even exploited by explorers. Both Ernest Shackleton and Captain Robert Scott took 'Forced March' cocaine tablets with them to Antarctica and during World War I it was supplied to some British troops to enhance their endurance.

Cocaine acts by preventing clearance of the neurotransmitter dopamine, released in response to nerve impulses, from the synaptic cleft. Consequently, dopamine hangs around longer and continues to stimulate its target cells. Amphetamine (speed) acts in a similar way. The addictive properties of both drugs come from the fact that dopamine stimulates the reward centre of the brain, so that pure cocaine produces feelings of exhilaration and euphoria, as Freud described. Providing that the body continues to be supplied with cocaine, dopamine levels in the brain remain elevated and the pleasurable sensation continues. When the drug wears off, however, the dopamine concentration plummets to below normal levels, producing depression, anxiety and a craving for more drug. Addiction, then, is an affliction of the brain and anything that stimulates the brain's reward centres to excess has the potential to be addictive.

Hooked

Nicotine is one of the most addictive drugs known. It is found in the leaves of the tobacco plant, *Nicotiana tabaccum*, which is named after the sixteenth-century adventurer Jean Nicot who brought the plant to France and is said to have popularized its use as a treatment for headache. Tobacco was introduced into England by Sir John Hawkins in 1565 and at first was met with amazement and considerable opposition. There is a famous story, probably apocryphal, that Sir Walter Raleigh's servant emptied a bucket of water over him in the mistaken belief that his master was on fire. Kings and papal bulls banned its use. King James I of England wrote a famous *Counterblaste to Tobacco* in 1604, calling it 'a custom loathsome to the eye, hateful to the nose, harmful to the brain, dangerous to the lungs and in the black stinking fume thereof, nearest resembling the horrible Stygian smoke of the pit that is bottomless'. Gradually, however, tobacco use proliferated, becoming widespread by the middle of the last century.

Smoking is an expensive habit in every sense. Every hour, twelve people in the UK die from smoking-related diseases and many more in the United States, and billions of dollars are spent on smoking-related health costs – over 190 billion dollars per year in the USA alone. It is estimated that half of cigarette smokers will eventually be killed by their habit, for smoking dramatically increases the risk of lung cancer (85 per cent of lung cancer cases are due to smoking) and is also associated with heart disease, stroke, emphysema and a range of other cancers. The link between smoking and lung cancer was established by Sir Richard Doll in the early 1950s, but the idea initially met with considerable resistance. Concerted health campaigns over the past fifty years have led to a decline in tobacco use and a corresponding fall in lung cancer rates, but around 20 per cent of adults still smoke. As everyone knows, it is not the nicotine in cigarettes that causes cancer but a cocktail of carcinogens contained in tobacco smoke; nicotine is dangerous because it is highly addictive and its tenacious hold makes it difficult to quit smoking.

Nicotine acts on acetylcholine receptors found at the junctions between nerve and skeletal muscle and at certain nerve–nerve synapses in the brain. Like acetylcholine itself, binding of nicotine to acetylcholine receptors opens an ion channel that allows sodium ions to enter the nerve cell and so stimulates it to fire off an electrical impulse. It is the drug's ability to activate certain brain neurones that accounts for its actions as a stimulant and, like caffeine, enables you to concentrate more effectively when tired. Its addictive properties stem from the fact that it also stimulates nerve cells in the reward pathways in the brain. Regular smokers adjust their smoking to maintain a constant concentration of nicotine in their blood and brain, and thus a steady level of neuronal stimulation. Some individuals have genetic differences in the liver enzyme that breaks down nicotine, so that the drug remains in their bloodstream for longer and they smoke fewer cigarettes to obtain the same effect.

Love, Love Me Do

'Tell me where is fancy bred, Or in the heart, or in the head?', asks Shakespeare in *The Merchant of Venice*. Romantic love has long been a favourite topic for authors, artists and playwrights, but what does cause us to fall in love? To seek the perfect mate and stick with them forever – or to favour the fresher faces, being constant only to inconstancy? The lovesick have often wished for a simple means to make the object of their affection fall in love with them. This is the basis of many charms and potions, perhaps even our predilection for cosmetics, perfume and dressing up. In *A Midsummer Night's Dream*, a love potion made from the juice of heartsease (the wild pansy) famously creates mayhem. It is administered to an unsuspecting Titania while she sleeps and compels her to fall in love with the first living thing she sees on waking. It happens to be Bottom, a most unprepossessing object of desire, since his head has been transformed into that of an ass. Yet although we may scoff at the idea of magical love potions, recent research suggests love is indeed no more than a chemical phenomenon.

Our understanding of the chemistry of attraction has its origin in a somewhat unlikely source – a rather unappealing small rodent known as the prairie vole. Prairie voles are monogamous and bond for life. In contrast, their cousins the montane voles are highly promiscuous, preferring to confine themselves to one-night stands. The difference in their behaviour seems to be related to two specific brain chemicals, oxytocin and vasopressin, that are released during mating. Oxytocin is crucial for pair bonding because its injection into the nucleus accumbens is sufficient to cement a couple for life, even if they are prevented from having sex. Conversely, if oxytocin action is blocked the prairie vole is only interested in fleeting affairs. Injecting oxytocin into a montane vole, however, does not dissuade it from a life of promiscuity, because it lacks oxytocin receptors in the reward centres of its brain. Oxytocin induces the release of dopamine and both are thought to act in concert to make pair bonding a particularly pleasurable experience. Vasopressin is similarly important for pair bonding, especially in males, and also provokes the aggressive behaviour that male voles display towards potential rivals during courtship and when defending the nest. It has also been linked to aggression in humans.

Clearly, it would be injudicious to extrapolate directly from voles to humans, for both the human brain and our social interactions are far more complex and involve multiple transmitters and many brain regions. Nevertheless, oxytocin is also important for bonding in humans. It is released during sex and suckling and may help cement the links between lovers, and between mother and child. It also enhances trust between people, an essential component of any loving relationship. Dopamine, that arbiter of pleasure and addiction, also plays an important role in romantic love. When the brain activity of students who claimed to be madly in love was examined in a scanner, dopamine-rich regions lit up when they were shown an image of their beloved. Thus here, to answer Shakespeare's question, is where fancy is bred, and in a very real sense, we may be addicted to the object of our love.

The (Un)Happiness Hormone

If pleasure is a construct of the brain, so too is misery. 'A mark in every face I meet, Marks of weakness, marks of woe' – as William Blake wrote, unhappiness is everywhere. So too is her more severe sister, clinical depression, for it is estimated that at some stage of our life almost 10 per cent of us will suffer from the 'black dog', as Winston Churchill termed it. In some people, it can be so severe that it is totally incapacitating.

Happiness and despair are the two faces of the neurotransmitter serotonin. Serotonin is produced by neurones of the raphe nucleus, whose processes ramify throughout the brain. Their targets include the nucleus accumbens and the ventral tegmental area, part of the brain's reward system. Because serotonin is released in many brain regions and interacts with at least fourteen different kinds of receptor, it affects many types of behaviour, but one of its most important roles is mood control. Elevated levels of serotonin are associated with feelings of optimism, contentment and serenity. Too little brings despair, depression, anxiety, apathy, and feelings of inadequacy. One way of increasing your serotonin levels is by vigorous exercise, which is why a brisk walk or a game of squash (if you can drag yourself off the sofa) helps relieve the blues. Modern antidepressants such as Prozac also act by elevating serotonin concentrations. They do so by inhibiting the removal of serotonin from the synaptic cleft, so that the transmitter stimulates its receptors for longer.

Perhaps the most infamous of drugs that interact with serotonin receptors, however, is LSD. Its extraordinary effects on perception are poetically described in the Beatles' song 'Lucy in the Sky with Diamonds'. But not all trips are so pleasant. In the TV series *Dr Who*, the Time Lord is 'regenerated' every few years, enabling him to be played by a different actor. Papers in the BBC archive explain to producers that the experience of regeneration is horrifying – it is, they say, 'as if he has had the LSD drug and instead of experiencing the kicks, he has the hell and dank horror which can be its effect'.

LSD is a psychedelic drug related to a natural compound, ergot-amine, found in the purple-brown fruiting bodies of the ergot fungus, *Claviceps purpurea*. It grows wild on rye and in mediaeval times contaminated rye bread caused dramatic outbreaks of ergot poisoning. Whole communities were sometimes affected. In 1930 the active ingredient of ergot was isolated and named lysergic acid, and subsequently the Swiss chemist Albert Hoffman produced a derivative that he named lysergic acid diethylamide – or LSD-25 for short. Although he did nothing with it for the next five years, he never forgot that experimental animals became restless when given the drug and, in 1943, having decided to reinvestigate the drug, he synthesized some more. Despite taking considerable precautions (for he knew ergot was toxic), during the final step of the synthesis he was overcome by a string of strange sensations including 'an uninterrupted stream of fantastic pictures, extraordinary shapes with intense, kaleidoscopic play of colors'.

Thinking this astonishing experience must have come from the drug, in the time-honoured tradition of pharmaceutical scientists, Hoffman cautiously ingested a tiny amount in a self-experiment three days later. It had a most dramatic effect. His notebook records, 'My surroundings had now transformed themselves in more terrifying ways. Everything in the room spun around, and the familiar objects and pieces of furniture assumed grotesque, threatening forms. They were in continuous motion, animated, as if driven by an inner restlessness. The lady next door, whom I scarcely recognized, brought me milk – in the course of the evening I drank more than two liters. She was no longer Mrs. R., but rather a malevolent, insidious witch with a colored mask.' Hoffman clearly had a bad trip, for he also remarks, 'A demon had invaded me, had taken possession of my body, mind, and soul.' He feared he was going to die, leaving his wife and three children bereft and his promising research work unfinished. Slowly, however, these horrors faded to be replaced by phantasmagorical visions of 'circles and spirals, exploding in colored fountains, rearranging and hybridizing themselves in constant flux', sounds that transformed themselves into optical images, and a sensation of renewed life.

LSD is one of the most powerful hallucinogens known. It has extraordinary effects on auditory and visual perception, producing a sparkling world in which colour, brightness and sounds are intensified, objects morph into strange shapes, and walls may 'breathe'. But altered percept and hallucinatory visions are not its only effects. It also produces changes in time perception, the emotions and self-awareness. Some users claim it even leads to higher states of consciousness (whatever those might be), spiritual awareness, even enlightenment. More prosaically, what all these experiences boil down to is changes in the electrical activity of the brain. LSD and other hallucinogens produce their 'magic' effects by binding very tightly to a specific subset of serotonin receptors at brain synapses, known as 5HT-2A receptors. Why LSD causes such intense hallucinations and serotonin does not, given that they both bind to the same receptors, is far from clear, but one clue may be that they seem to trigger different signalling pathways in their target cells.

The Art of Memory

Our understanding of the physiological basis of emotions other than pleasure and despair – of anger, embarrassment, envy, grief, disgust, guilt and astonishment, to name but a few – is less clearly established. What is well recognized, however, is that our emotional reaction to a given situation is strongly influenced by our previous experience. Memory plays a key role in how we feel, and it is in the amygdala, two almond-shaped brain regions that lie on either side of the head, that memories are interwoven with emotions. Here, too, reward and fear memories are stored and recovered.

Memory – how to enhance it, and how memories are laid down, stored and retrieved – has perplexed and fascinated people for centuries. In the days before cheap paper, or computers, it was of particular importance. The Ancient Greeks and Romans were especially skilled in the art of memory, for lawyers and politicians were expected to speak for hours without notes. Consequently, methods of remembering were widely discussed. Quintilian tells of how the poet Simonides delivered the victory ode for his host, a champion boxer, at a magnificent banquet in Thessaly. As was traditional, his panegyric included a passage that lauded the twin gods Castor and Pollux. Annoyed at having to share the credit, and despite the price having been agreed beforehand, Simonides' host withheld part of the fee, telling him he should claim the balance from Castor and Pollux. A little later, Simonides was summoned from the room by a message that two young men wished to see him urgently. Scarcely had he left the building before it collapsed and all inside were crushed to death beneath the rubble. The callers who had saved Simonides's life had vanished, but were assumed to be Castor and Pollux. The message of this story was not, as one might imagine, the moral importance of paying one's bills, but rather of what Simonides did next. So badly mutilated were the bodies of the dead that it was impossible to recognize any of them. However, Simonides was able to remember the precise positions in which all the diners had been sitting, thus enabling their bodies to be restored to their respective families. He had invented the 'art of memory'.

Referring to the story, Quintilian recommends that when learning a long text you should break it up into shorter pieces. Then you should visualize a familiar place – your home, for example – and put different bits of the text in different rooms. To recall the text again, you just walk through the imaginary house, room by room, recollecting the text as you go. The place method, and continual repetition, are still the best ways to remember something and are often used by memory savants today.

Remembrance of Things Past

Exactly how and where memories are stored in the brain is still unclear. The fact that stimulation of certain bits of the brain can evoke vivid memories of things past – a familiar scent, a snatch of a song, even the complete recall of an event with all its sensations intact – suggests that at least some memories are stored in certain specific brain locations. People who suffer from visual agnosia may lose the ability to recognize particular objects, despite their senses and memory being intact. As Oliver Sacks relates in his book *The Man Who Mistook His Wife for a Hat*, they can describe what a glove looks like, but may be quite unable to recognize that it is, in fact, a glove, or know what it is used for. They may also fail to recognize one person, but not another, or confuse their wife with their hat. All this suggests that there may be discrete regions of the brain that are used for processing and storing specific types of information.

There is also a distinction between short-term and long-term memory. You use the former when you remember a number for a few minutes, or plan the outcome of a series of chess moves before you decide which one to use. Short-term (or working) memory seems to involve regions within the cortex, particularly the frontal lobes. Long-term memory enables us to recall events from our childhood. The fact that most people fail to recollect many events before the age of about three suggests that long-term memory storage may not be fully developed until then. How working memories are selected for long-term storage, how, where and in what form they are laid down, and how they are retrieved is currently under intensive investigation.

One brain region of key importance for memory storage is the hippocampus, so called because it is shaped like a seahorse. We have two of them, one on each side of the brain. Their role was discovered serendipitously by studies of Henry Gustav Molaison, better known to scientists as HM. As a young boy, HM suffered from intractable epilepsy. In an attempt to cure his seizures, most of the hippocampus on both sides of his brain was removed when he was twenty-seven. The consequence was disastrous for HM (but a goldmine for science), as he lost the ability to make new memories and his memory of some preceding events was also impaired. He was confined to living in the past. Nevertheless, he was able to perform tasks that need only short-term memory, clearly demonstrating that short-term and long-term memory are distinct. His ability to learn new motor skills was also intact; he became an accomplished table-tennis player, despite being adamant he had never played it before. He was also a gracious, patient and modest individual whom the researchers he worked with considered as one of the family, although he never recognized who they were, even if they returned just a few minutes after leaving the room.

The hippocampus is particularly important for spatial memory – for our ability to recall places. Taxi drivers who have to memorize the streets of London, information colloquially known as the 'Knowledge', tend to have larger hippocampi than the rest of us. Brain imaging has revealed that their hippocampi also light up when they are planning a route; simply thinking about how they might travel from Paddington Station to Buckingham Palace activates this bit of the brain. Fascinatingly, when they cease to use the 'Knowledge' regularly their hippocampi revert back to the same size as ours. 'Lose it or use it' seems to be an aphorism that is as valid for the brain as for your muscles.

It turns out that we construct a spatial map of our environment inside our heads, which can even be detected at the level of single hippocampal neurones. Nerve cells known as 'place cells' increase their activity only when an animal is in a specific location in its environment. As a rat runs along a corridor, for example, an individual cell bursts into activity and then ceases to fire as the animal enters and leaves the location corresponding to its 'place field.' Multiple neurones, each with a different place field, together provide an 'electrical map' of the whole environment. This map is established within minutes of entering a new environment, and if the animal is returned to the same environment a few days later, the same nerve cells fire at exactly the same location. Thus this spatial reference map may be involved in formation of spatial memories.

Although the hippocampus is crucially involved in laying down long-term memories, most memories are not actually stored there. Many other bits of the brain appear to be involved. Imagine you are watching an opera – *The Magic Flute*, for example. Your eyes capture the image of the Queen of the Night robed in a gorgeously coloured gown, and your ears pick up the wonderful aria she sings. These are relayed to the visual and auditory cortex respectively, where they are interpreted, and linked together to create a picture of the scene. The information is then forwarded to the hippocampus, which decides whether to pass it on to your long-term memory. If it does, the information is relayed back to the appropriate cortical areas, where it is laid down as new synapses or existing connections are strengthened. Information is thus circulated around the brain; it is not a matter of it being channelled straight from the eyes to your memory, but of a complex series of information-processing events that take place in multiple different brain regions.

The hippocampus enables associations between sensations and experiences to be hard-wired, enabling you to 'play back' a scene from memory. Damage to this bit of the brain affects your ability to store new memories. However, memory formation does not only involve the hippocampus. The amygdala also plays a part in memory consolidation. How interested you are in an event and what emotional associations it has for you will influence your ability to recall it later. This is why most of us will remember events such as the birth of our child or where we were when we heard that the Twin Towers of the World Trade Center had collapsed, but will probably quickly forget what we were doing at lunchtime last Tuesday.

Memories of mechanical skills are stored separately and are not channelled through the hippocampus. Your ability to remember how to ride a bike even though you have not done so for many years is stored in your cerebellum and motor cortex. This is why it is still possible for people to play music despite having lost much of their memory of places and events, and why HM was able to play table tennis.

Memories are Made of This

How memories are laid down appears to involve changes in the physical structure of the brain. Contrary to what was once believed, your brain is not a static structure, but extraordinarily adaptable. New connections between nerve cells are continuously being made and existing ones strengthened or eliminated as you go about your daily life. This process, known as synaptic plasticity, is the physical basis of learning and memory.

The fine filaments – the dendrites – that extend outwards from the nerve cells in your brain are covered in tiny knob-like extensions called spines. Thousands of them decorate a single dendrite. The dendritic spines are the sites of the synapses and it is here that memories are hard-wired, for as we learn new things and lay down new memories, new spines appear and existing ones change shape or disappear. As they grow in size and number so a particular neuronal pathway is reinforced. Such reinforcement often happens when connections between neurones are simultaneously activated, and has given rise to the neuroscientists' adage 'Cells that fire together, wire together.' This happens very rapidly. Experiments in mice have shown that learning to press a button to obtain a food reward is associated with a dramatic increase in new synapses within just an hour of starting training. Strikingly, in these experiments, the new spines endured long after training had ceased, but the total number gradually returned to the pre-training level because older ones were eliminated. Perhaps the brain can only support a finite number of connections, which is why learning new things may reduce our capacity to remember older events. Ion channels lie at the heart of memory for the presence of different kinds of glutamate receptor channels is necessary both to retain existing synapses and to grow new ones. If these channels are absent, or their function is impaired, our ability to remember is reduced.

As we grow older our remembrance of things past often seems to diminish. The semi-photographic memory I had as a child has long since vanished and my ability to recall names and faces is now embarrassingly poor. But this is nothing compared with the trauma of Alzheimer's disease, which afflicts around half a million people in the UK. It is the most frightening of diseases for it steals away the soul. At first it may seem as if the victim has no more than mild memory loss, but with time they lose all recollection of friends and family, become confused, withdrawn and increasingly distressed.

Alzheimer's disease is characterized by the loss of neurones and synaptic connections in the cerebral cortex, resulting in a reduction in the size of the brain. Tangled networks of a protein known as tau appear inside nerve cells and dense plaques of amyloid protein are found in the space between nerve cells. Whether these are the cause or the consequence of cell death is unknown. The electrical activity of the brain is clearly impaired, but again whether this is merely due to the loss of nerve cells, is produced by the observed reduction in dendritic spines, or results from impaired transmission between nerve cells is unclear. One idea is that the disease leads to a reduction in the amount of acetylcholine in certain regions of the brain, and thus drugs that block the breakdown of the transmitter are currently used as therapy to boost its levels. They are not very effective, however, merely slowing the progression of the disease. Nothing has been found that can stop it in its tracks or reverse its effects. At present, Alzheimer's disease has no cure and remains a tragedy for both the patient and their family.

Shedding Light on Behaviour

Understanding exactly how the brain controls a particular behaviour is far from easy. One approach has been to try to tease apart the precise contributions of individual neurones. Recent pioneering work by Oxford University professor Gero Miesenböck has led to a revolutionary new field of neuroscience called optogenetics which enables a particular group of nerve cells to be turned on (or off) at will, without affecting the activity of adjacent neurones. In this way, it is possible to control the behaviour of an animal simply by switching on a light. The technique utilizes ion channels that act as light-activated molecular switches. These are inserted into a specific set of nerve cells by genetic manipulation, where they sit quietly shut, without any effect on the cells' electrical activity, until the researcher chooses to open them by illuminating them with an intense pulse of laser light of a particular wavelength. One of these light-activated ion channels, known as channel rhodopsin, comes from a green alga. Simply switching on the laser light opens the channel, leading to an influx of positively charged ions that stimulates the cell into activity. Because the duration and timing of the laser pulse can be precisely controlled it is possible to mimic the activity of individual nerve cells and thus investigate how different patterns of activity influence behaviour. In a similar fashion it is possible to turn off the electrical activity of a nerve cell using a different kind of light-activated ion channel that clamps the cell at the resting membrane potential when it is opened.

To woo a mate, the male fruit fly sings to her by rapidly vibrating his wings. Miesenböck was fascinated by the fact that although the brains of male and female flies seem to be wired up in much the same way, their behaviour is very different. His team found that by switching on a specific group of neurones with a light pulse, female flies could be coaxed into producing the male courtship song. It is as if the fruit fly has a 'unisex' brain that is directed to produce different patterns of behaviour – male or female – by a few neuronal master switches. If the correct nerve cells are stimulated, a fly can even 'learn' from an experience it has never had. While it is relatively easy to control a fruit fly's behaviour with light, it is more difficult to do so in a mammal, as the laser beam cannot penetrate the skull and light must be delivered by a fibre-optic cable implanted in the brain. Nevertheless, it has also proved possible to control the behaviour of a mouse this way. Optogenetics promises to be a valuable tool for illuminating how the brain controls behaviour.

Just as the fruit fly's courtship song-and-dance routine is hard-wired, so too are other forms of social behaviour. Moreover, experience physically shapes our brains, which helps explain why identical twins, despite having exactly the same genetic constitution, are quite different people. This is beautifully illustrated by the social hierarchy of the crayfish. When challenged, a crayfish will back out of a potentially threatening situation using a tail flick that catapults it rapidly backwards. If two crayfish are placed in the same tank, one quickly becomes dominant and the other subordinate, and this is paralleled by a marked difference in the electrical responses of the giant nerve fibres that control the tail flip, and in the effect that the neurotransmitter serotonin has on these cells. If the dominant animal is removed from the tank, the subordinate one then adopts a dominant electrical response by changing the way in which serotonin acts on its nerve cells. Fascinatingly, once a crayfish has experienced being the alpha animal for a while it never looks back. Although it may revert to subordinate behaviour if a more aggressive crayfish is reintroduced into the tank and it loses a fight, the 'dominant' effect of serotonin remains unchanged. In a kind of neurological denial it has forever a dominant brain. Reality television programmes in which individuals play different roles in Edwardian society reveal that people quickly assume servant or master roles. An interesting question is the extent to which such role-playing may have physically changed their brains.

To Sleep, Perchance to Dream

Sleep is so familiar that we rarely think about it. Every night when we fall asleep we surrender our consciousness, our muscles relax and our ability to respond to mild stimuli is diminished. Sleep is associated with characteristic changes in the electrical activity of our brain, but this is not simply a global suppression of nerve cell function but a highly controlled phenomenon. Although we commonly think of sleep as a single state it actually comprises two quite distinct brain states, known as rapid eye movement (REM) sleep and non-rapid eye movement (NREM) sleep. Throughout the night, periods of REM sleep alternate with those of NREM sleep. Each of these sleep cycles lasts about ninety minutes and you will have about four or five of them a night. In total, around 25 per cent of the time you spend sleeping, amounting to between one and a half to two hours a night, is spent in REM sleep, with its duration increasing in each sleep cycle as night moves towards morning.

As you fall asleep, you first enter a transient dream-like state between sleeping and waking. This twilight zone lasts just a few minutes, after which you enter a period of NREM sleep. As you do so, your EEG pattern changes, passing through various stages of light sleep before finally settling down to the slow, rolling, low-frequency brain waves of deep sleep. Your muscles relax and your ability to respond to external stimuli diminishes. Brain activity in many areas, particularly the cerebral cortex, is reduced and you are now hard to wake.

Astonishingly, after you have been sleeping for about an hour or so everything abruptly changes. Despite remaining sound asleep, your brain appears to wake up and your EEG becomes a frenzy of rapid low-voltage, high-frequency waves. Many areas of your brain become activated, with particularly intense activity occurring in regions associated with the emotions, such as the amygdala. This is a time of intense dreaming and if you are woken you are likely to remember your dreams. Your muscles are paralysed by inhibitory signals sent from the brainstem to the muscles to prevent you damaging yourself by acting out your dreams. The only muscles that remain active are your respiratory muscles (fortunately) and those of your eyes, which are connected directly to your brain and therefore bypass the inhibitory pathways in your brainstem. Flurries of rapid eye flickers occur, which is why this stage of sleep is known as rapid eye movement (REM) sleep. If the brainstem is damaged, the ability to inhibit muscle movements during sleep may be lost and such unfortunate people may get out of bed and move about during their dreams: they may even need physically restraining to prevent them from hurting themselves or their sleeping partners. When you exit REM sleep, muscular control is automatically re-engaged. In some rare individuals this does not happen immediately and they may wake to find themselves temporarily paralysed, a truly frightening experience.

During REM sleep your senses are also disconnected from your brain so that you are cut off from the world. The brain region known as the thalamus relays sensory information from our sense organs up to the cerebral cortex, but during sleep this pathway is largely closed, so little gets through. We are walled off from the world, in sensory isolation and unable to command the use of our muscles, but our brains are on overdrive. It is rather like a car whose engine is revving, but which cannot move because the gears are not engaged. Sleep, then, is a dynamic activity. Your brain does not simply switch off, but instead refocuses its activity differently.

At some time of our life, we have all felt unaccountably sleepy during the day – often as a consequence of jet-lag – but for some unfortunate people excessive sleepiness is more of a problem. Take Claire, for example, who abruptly and embarrassingly falls asleep at inopportune moments and is quite unable to do anything about it. She also once laughed so much at her friend's joke that her legs gave way and she collapsed on the floor; indeed any form of excessive excitement or strong emotion can cause a loss of muscle tone so profound that she becomes floppy and falls over. Claire suffers from a chronic sleep disorder known as narcolepsy that is characterized by excessive daytime sleepiness despite adequate night-time sleep. The lack of muscle control she experiences is known as cataplexy and arises because the inhibitory pathways that prevent us moving during sleep are inappropriately switched on during wakefulness or very early in the sleep cycle.

A group of scientists led by Emmanuel Mignot investigated a strain of Doberman pinscher dogs that suffered from the same condition as Claire. Spotting a special food treat, one of these dogs will rush happily over and take a few mouthfuls, but after a few seconds will be so overcome by the excitement that it loses control of its limbs and collapses. The condition is inherited and by laboriously searching for the gene involved scientists identified a chemical called orexin (or hypocretin) that helps stop us falling asleep. This is produced by a small region of our hypothalamus and keeps us awake by stimulating electrical activity in other regions of the brain. If you lack the ability to make orexins, or you have insufficient orexin receptors (as the Doberman dogs did), you will fall asleep involuntarily.

Sleep is a universal imperative. All animals sleep, even insects and fish, and while we proverbially manage an average of eight hours a night, some creatures sleep for far longer. The champion is the two-toed sloth, which sleeps as much as twenty hours each day. Mammals that live in the sea would drown if they fell sleep underwater, so they rest half of their brain at a time, with one side remaining awake while the other is deeply asleep. So too do many birds, which often sleep away the night with one eye open, keeping watch for predators.

Quite why we sleep is still something of a mystery, but there is evidence that one reason is that it is important for memory consolidation. As you will no doubt already know from experience, without adequate sleep our ability to remember things diminishes. Strikingly, even a short nap can help with learning a new task. One hypothesis is that while we are laying down long-term memories, and consolidating and organizing new knowledge, it is important that there is no new input to confuse things. Cut off from the outside world, memories can be replayed, strengthened, stored or discarded more easily. Some evidence in favour of this view comes from the finding that during sleep hippocampal 'place cells' fire in a coordinated fashion that suggests the spatial reference map that is formed in the brain when an animal is exposed to a new environment is being replayed. It is as if the brain is remembering its earlier experience – although whether this replay is associated with the consolidation of memory or the transfer of memories out of the hippocampus to areas of the brain where memories are stored is still unknown.

Whatever its actual function, sleep is essential, for without it we soon die. Sleep, then, is very far from the 'little slice of death' that Edgar Allan Poe bewailed. Nor is it a nightly waste of time. Rather, as Shakespeare put it, it is 'the chief nourisher in life's feast' and we should endeavour to make sure we get enough.

The God of Dreams

Once, when I was a child, my whole family came down with severe gastroenteritis. Both my parents and my siblings were hors de combat and I was the only person available to collect the medicine. We lived in an isolated village and there were no buses running at the time, so I jumped on my bicycle and pedalled the five miles to the doctor's surgery. I was given a large glass bottle of kaolin and morphine that I stuffed in my jacket pocket. The journey back was something of a nightmare as it soon became clear that I too was suffering from the vomiting bug and I had to stop several times to throw up on the verge. I arrived home completely exhausted. My mother immediately gave me a large dose of medicine, but, being ill herself, she did not think to shake the bottle first. During the long journey, the kaolin had settled to the bottom and floating at the top was a (weak) solution of pure morphine. Morphine, named after Morpheus, the Greek god of sleep, is a powerful sedative and I slept for twenty-four hours.

Morphine is not only a sedative. It also produces relaxation, induces a state of delightful, dreamy euphoria and has the great virtue of relieving pain. It has been used for thousands of years, as both a medicinal and a recreational drug, in the form of opium, a crude extract of the opium poppy, *Papaver somniferum*. As the seventeenth-century physician Thomas Sydenham said, 'among the remedies it has pleased Almighty God to give to man to relieve his sufferings, none is so universal and so efficacious as opium'. Historically, a mixture of opium and alcohol called laudanum was used to treat a variety of ailments and as a consequence many people became addicted, including the poet Samuel Coleridge, who is believed to have written 'Kubla Khan' while under the influence of the drug. Others took it for pleasure. Thomas de Quincey wrote in his famous *Confessions of an English Opium Eater*, 'I, wretch that I am, being so notoriously charmed by fairies against pain, must have resorted to opium in the abominable character of an adventurous voluptuary, angling in all streams for a variety of pleasures.'

Like other addictive drugs, morphine stimulates the reward centres of the brain, and when taken to excess produces feelings of pleasure so intense that when the 'high' wears off the addict craves for more. Its use has led to more than individual misery. By the late eighteenth century, a lucrative and cynical trade had grown up between Britain and China, brokered by the East India Company. Tea, at that time produced only in China, was in high demand in Britain. China demanded payment in silver, but as this was expensive British merchants gradually switched to trading opium for tea. It was trafficked in surreptitiously from India by an indirect route. This proved disastrous for China, as its addictive nature created an instant demand for opium and much of the peasantry and army became incapable of work or combat as a consequence. When China prohibited its sale and requested the illegal imports be halted, Britain, mindful of the economic value of the trade, refused. The issue escalated into the infamous Opium Wars, led to Hong Kong being ceded to the British, the establishment of tea plantations in

India, and new trade treaties. Tea, in its own way, appears to be almost as addictive as opium (at least to the British).

Morphine and heroin, to which it is structurally similar, belong to a class of drugs known as opiates. They bind to opiate receptors in the brain and spinal cord, thereby shutting calcium channels and inhibiting transmitter release and neuronal electrical activity. It was musing about why we actually have opiate receptors in the first place that caused John Hughes and Hans Kosterlitz to speculate that the body might produce its own opiates. That led to the discovery of endorphins (the term derives from *endo*genous mor*phine*), chemicals produced by the body that are released in response to pain. They are also the chemicals that produce the feeling of wellbeing you get from running and other forms of intensive exercise.

Like synthetic opiates, endorphins suppress electrical impulses in pain nerve cells. If you have ever seen a twitch put on a horse you will know just how powerful endorphins can be. The twitch is a loop of rope that is twisted around the animal's sensitive upper lip and is often used to quieten a restive horse while it is being shod or examined. As the rope – and the lip – is twisted and tightened, the horse appears to almost fall asleep: it stops dancing around, its head droops, its eyes glaze over and it stands quietly. It does so because its system is flooded with endorphins, released in response to the intense pain produced by twisting its lip. Acupuncture also triggers endorphin release, which may explain why some surgical operations can be performed without anaesthesia.

Knockout Drops

Humphrey Davy was a self-experimenter extraordinaire. Undeterred by almost killing himself by breathing pure carbon monoxide, he went on to investigate the physiological effects of many other substances. Around 1799, he discovered that nitrous oxide – previously believed by some to be a lethal gas – induced euphoria and uncontrollable outbursts of laughter, and led him to dance around his laboratory like a madman. He christened it laughing gas. Inhaling nitrous oxide soon became part of his daily routine. Strangely, however, despite the fact that Davy recognized that the gas took away the sensation of pain and even induced loss of consciousness he never seemed to appreciate its potential as an anaesthetic.

In yet another example of how a British discovery was subsequently exploited to great effect by entrepreneurs in the United States, it was a group of US dentists, searching for a means of extracting teeth painlessly, who introduced general anaesthetics for pain relief. One of the first to do so was Horace Wells, who trialled nitrous oxide first on himself, and then on his patients. Confident of its virtue, he gave a public demonstration of a tooth extraction under ether anaesthesia to a class of medical students in Boston in 1845. It was not a success, as the gasbag was removed too soon, the patient yelped with pain and the watching students jeered and booed. Wells was so disheartened by the affair that he gave up dentistry, became addicted to chloroform, and threw sulphuric acid at two prostitutes while under the influence of the drug. When he realized what he had done, he committed suicide. Wells's misery was compounded by the fact that his colleague, William Thomas Green Morton, performed the first successful public operation on a patient under ether anaesthesia in Boston just a year after his own abortive attempt. Unlike Wells, Morton was widely lauded. He was far less popular, however, for his attempts to make money from the process: his patenting of the use of ether as an anaesthetic caused a public outcry and his demand that Congress pay him 100,000 dollars as a 'national recompense' for his invention was met with scorn.

Because ether tends to irritate the lungs (and explodes easily), James Simpson, a Scottish obstetrician, subsequently introduced the use of chloroform to ease the pain of labour. Although some opposed the practice, citing the teachings of Genesis in which God tells Eve, 'in sorrow thou shalt bring forth children', its use received a considerable boost when it was administered to Queen Victoria on 7 April 1853 during the delivery of her eighth child, Prince Leopold. The *British Medical Journal* commented, 'the Royal

Majesty of the patient, and the excellence of her recovery, are circumstances which will probably remove much of the lingering professional and popular prejudice against the use of anaesthesia in midwifery'. Fascinatingly, the Court Circular reports that during the birth the Queen was attended not only by her doctors and a nurse, but also by Prince Albert, proving him to have been a thoroughly modern husband.

A good general anaesthetic needs to induce loss of consciousness, loss of pain sensation (analgesia), immobility and preferably also amnesia so that you fail to remember the experience. All of this must be achieved without affecting the heart and, if possible, the respiratory muscles, and without causing vomiting or long-lasting neurological complications. And of course it must be easily reversible. Not all of these attributes are necessarily found in a single drug. Ether and chloroform are effective anaesthetics, but they are far from perfect, and today vapours such as isoflurane and sevoflurane, or injected drugs like propofol, are usually used to induce loss of consciousness and amnesia, with other drugs being used to produce analgesia and muscle relaxation. Interestingly, nitrous oxide gas still has a useful role in many cases.

As a general anaesthetic takes effect, you first enter a sedated state: you become drowsy and even if the anaesthetic is removed you will remember little of the experience. Gradually, you fail to respond to verbal commands and drift into unconsciousness. Surprising as it may seem, you then enter an excited state characterized by uncontrolled movements and irregular breathing. Subsequently your muscles relax again, breathing becomes regular, eye movements cease and you are now so fully asleep that a surgeon can perform an operation without causing pain. At this stage, brain scans show that metabolic activity is uniformly suppressed across your brain, suggesting that all brain regions are affected. It seems there is no brain region that is especially sensitive to general anaesthetics. Frustratingly for neuroscientists trying to identify where consciousness arises in the brain, it has not been possible to identify an area whose activity 'winks out' just as consciousness is lost.

Despite the fact that general anaesthetics are administered to thousands of people throughout the world every day, we still have only a limited idea of how they work, and how they induce unconsciousness remains one of the great mysteries of neuroscience. Current evidence suggests that they suppress brain electrical activity by interacting with ion channels. It is suggested that they stabilize a particular conformational state of the channel by occupying gaps in the protein molecule itself or by intercalating between the protein and the lipid membrane in which it sits. Some anaesthetics seem to open ion channels that suppress the electrical activity of brain cells, such as GABA channels, glycine receptors and potassium channels. Other anaesthetics block synaptic transmission by inhibiting the function of excitatory glutamate receptors. The fact that both excitatory and inhibitory neurones are affected fits with the general suppression of electrical activity seen in brain scans.

Who Am I?

Precisely what consciousness is has occupied philosophers and neuroscientists for centuries and we still lack a definitive understanding. Yet it is something that each of us is so familiar with and that we all experience. 'I think', said René Descartes back in the fifteenth century, 'therefore I am'. But what, exactly, am 'I'?

In Descartes's view, the mind and body were separate entities. But the profound changes in our personalities produced by drugs, disease and brain damage provide abundant evidence that this is not the case – our minds are the product of our brains. Parenthetically, I have often wondered if the Cartesian view may in part stem from the fact that philosophy was once primarily the preserve of men, for the penis often appears to have a life of its own, sometimes refusing to perform when desired or being embarrassingly eager at inopportune moments. Women, on the other hand, whose emotional state is clearly influenced by their hormonal cycle, are constantly reminded of the link between body and mind.

Despite our very powerful sense of self, neuroscience reveals we are no more than the integrated electrical activity of our brain cells. Uncomfortable as it may seem, there is no separate entity, no soul, and nothing that lives on after our death – a fact that catapults science into direct conflict with many religions. So where does it come from, this precarious feeling of 'I', this person sitting here inside my head, looking out of my eyes, tapping away at this keyboard, trying to communicate my thoughts to you?

We are not born with a sense of self. Babies are not self-aware, nor do they recognize the thoughts and feelings of others. They develop these attributes gradually. A common way to assess self-awareness in very young children or animals is to test their ability to recognize themselves in a mirror by sticking a brightly coloured label on their head. If they identify themselves they will try and remove the label – but if they see a stranger they will do nothing, or reach out towards the image in the mirror. By this criterion, human children become self-aware between the age of two and three. It seems that our brains must reach a certain stage of development before we are fully self-aware. It is also not an instant 'awakening' – psychologists suggest that there are several steps in the development of self-awareness.

The next obvious question is where self-awareness is located in the brain. Is it a distributed entity, involving multiple networks of nerve cells, or does it reside in a specific set of nerve cells? Studies of brain-damaged patients suggest there is no discrete site for self-awareness, because while damage to specific regions of the brain can cause dramatic changes in personality, it does not create a zombie – an individual with no sense of self but otherwise functionally normal. Likewise, no one brain region 'winks out' concomitantly with the loss of consciousness when you are given a general anaesthetic. The loss of consciousness we experience during sleep is also very different from that of anaesthesia; anaesthesia seems to cause a general depression of electrical activity across the brain, but sleep is an exquisitely regulated active condition. The very different patterns of brain activity during sleep, anaesthesia and wakefulness indicate that loss of consciousness – and by implication perhaps also consciousness itself – may have more than one origin. It also seems to require the integrated activity of many neurones – but exactly which ones is still a mystery.

It is also worth noting that memory is intimately connected with our perception of consciousness. Patients given certain sedatives are able to respond to the doctor's commands, can feel pain and would probably claim to be conscious if asked. Yet when the drug wears off they remember nothing and will state they were unconscious throughout the operation. A similar phenomenon is found with notorious date-rape drugs such as rohypnol, which can produce profound amnesia. Perhaps, then, one reason that children do not develop self-awareness until the age of two or three is that long-term memories do not appear to be laid down until about the same age – few of us remember anything before we were three.

And yet, you may argue, it feels so real, this individual inside my head. How can it be no more than an illusion? But remember that we are easily fooled. Our brain shapes the way we perceive the world, and the way we react to it. It regularly seduces us into thinking we see or hear something other than we do, for visual illusions abound and attention may wax and wane. Brain imaging even indicates that we may act on a decision, such as whether to press a button with our right or left hand, even before we are aware of having made it. Free will, like so much else, is merely an illusion. It, too, is a construct of our brain.

There must be many reasons why humans have evolved consciousness, but perhaps one is that self-awareness is linked with our ability to appreciate the thoughts and feelings of others. This is crucial for teamwork and social cohesion, attributes that have been critical for the success of our species. The only other creatures to evidence self-awareness in the mirror test are also social animals – they include chimpanzees, elephants and dolphins.

The origin of consciousness is one of the most challenging questions of our time, and something of a minefield for philosophers and neuroscientists alike. It is far too complex to tackle fully here, in just a few lines. Nevertheless, most scientists would now agree that consciousness emerges from the electrical activity of the brain – and that, in turn, derives from the activity of my favourite proteins, the ion channels. And although there remains a huge gulf in our understanding of precisely how neuronal activity shapes cognitive function, new technologies promise that we may eventually understand how behaviour is generated and regulated. They may also provide new insight into the origin of thought and feelings. How our minds work is no longer the province of philosophers and theologians. It is now the subject of neuroscience. For our thoughts and emotions, our feelings of self, reflect a maelstrom of electrical signals whirling around the brain. Mary Shelley was closer to the truth than she perhaps appreciated when she inferred that electricity is the spark of life. We are indeed no more than electrified clay.

12

Shocking Treatment

Your temples, where the hair crowded in,
Were the tender place. Once to check
I dropped a file across the electrodes
Of a twelve-volt battery – it exploded
Like a grenade. Somebody wired you up.
Somebody pushed the lever. They crashed
The thunderbolt into your skull.
In their bleached coats, with blenched faces
They hovered again
To see how you were, in your straps.

Ted Hughes, 'The Tender Place'

On the fourth of January 1903, a state execution was held at Lunar Park on Coney Island in New York city. It was witnessed by more than 1,500 'curious persons' who had flocked there to see the show. The victim was a twenty-eight-year-old elephant named Topsy. Topsy had once toured the United States with the Forepaugh Circus, but ended her days at Coney Island, where she had become violent and aggressive and killed three zookeepers. On at least one of these occasions she had been severely provoked, for her keeper had maliciously tried to feed her a lighted cigarette. He was in for a surprise, as she grabbed him with her trunk and threw him to the ground, killing him instantly. Her owners decided she was too dangerous to keep, but dispatching a six-ton, ten-foot-tall elephant is far from easy. Poison proved a failure. Thomas Edison, who was locked into a battle over the virtues of AC and DC electrical systems, saw a publicity opportunity and suggested she be electrocuted. So Topsy was fitted with copper-lined sandals, anchored in place with chains and given a massive electric shock of 6,000 volts. Death was immediate. Edison filmed the unedifying spectacle with a motion-picture camera he had made and exhibited it to audiences round the country to demonstrate the dangers of AC current.[1] It was, as the *New York Times* said at the time, 'a rather inglorious affair'.

Electricity not only powers our bodies, it can also be used to manipulate them. This final chapter considers how electricity has been used for good and ill. It charts the use of electricity in medicine from classical times, through its scientific origin in the late eighteenth and early nineteenth centuries, to the present day, and reflects on how it has the potential to transform our lives even further. It also considers the darker uses of electricity – to kill, maim or control others.

Electricity Made Plain and Useful

The medicinal use of electricity has its origins in antiquity. As long ago as 46 AD, the Roman physician Scribonius Largus recommended the use of the electric ray *Torpedo* as a cure for the pain caused by gout and headaches. Headache, he wrote, 'even if it is chronic and unbearable, is taken away and remedied by a live torpedo placed on the spot which is in pain, until the pain ceases'. This induced a numbness that dulled the pain. Scribonius further advised that several fish should be prepared, as two or three were sometimes needed to effect a cure. For gout, 'a live black torpedo should, when the pain begins, be placed under the feet. The patient must stand on a moist shore washed by the sea and he should stay like this until his whole foot and leg up to the knee is numb. This takes away present pain and prevents pain from coming on if it has not already arisen.' Torpedoes are not always easy to come by, their effectiveness is limited as they soon die when removed from water, and hanging around on the shoreline is decidedly inconvenient. Thus it was not until the invention of electrostatic generators in the eighteenth century that electrotherapy became widespread.

One pioneer of electric shock treatment was the preacher John Wesley, the founder of the Methodist movement. Wesley became interested in electricity in the late 1740s as a result of attending several public demonstrations of electrical fire and reading of Benjamin Franklin's experiments. Moved by the plight of his flock, most of whom had little access to medicine, he saw electricity as a 'shockingly cheap and easy way' to treat a multitude of ailments. Around 1753, he procured an electrical machine and experimented with it both on himself and on others. Its beneficial effects so impressed him that he introduced electric shock machines at free medical clinics in Bristol and London, where the poor could be electrified.

He wrote, 'I ordered several persons to be electrified, who were ill of various disorders; some of whom found an immediate, some a gradual cure. From this time I appointed, first some hours in every week, and afterward an hour in every day, wherein any that desired it might try the virtue of this surprising medicine. Two or three years after, our patients were so numerous that we were obliged to divide them: so part were electrified in Southwark, part at the Foundery, others near St. Paul's, and the rest near the Seven Dials: the same method we have taken ever since; and to this day, while hundreds, perhaps thousands, have received unspeakable good, I have not known one man, woman, or child, who has received hurt thereby.' Based in part upon his successful results, in 1760 he published *Desideratum*, a treatise on 'electricity made plain and useful by a lover of mankind and of common sense'. In it he records many examples of muscle cramps, headaches and rheumatism that were apparently healed by electrotherapy. Interestingly, he felt that electrotherapy was of special benefit in nervous disorders.

The machine Wesley used appears to have been a simple electrostatic generator.[2] Turning a handle caused a glass cylinder to rub against a piece of silk, generating a static charge that was collected by comb-like spikes attached to a thin metal rod, which the patient presumably grasped. Alternatively, the charge generated could be drawn off and stored in a Leyden jar. The magnitude of the shock could be controlled by the number of times the handle

Not all patients appeared to be completely happy with the experience of electrotherapy.

of the generator was cranked or, in the case of the Leyden jar, by the size of the jar (smaller jars produce lower jolts of electricity). The doses Wesley used seem to have been based in part upon the work of Richard Lovett, a lay clerk at Worcester Cathedral, and in part on his own experiments and observations. The mild shock felt was seen as the passage of 'electrical fire' through the body and was presumed to be of medical benefit. The shocks Wesley used were fairly mild and are likely to have been safe, but it is far from clear whether the treatment was really effective. Even today there is debate about whether mild electrical stimulation can help in the treatment of muscle atrophy and chronic pain, and it seems more likely that any benefit people received was due to the placebo effect and the sympathetic concern of the practitioners.

Wesley's attempts to heal the poor were not met with universal acclaim. Some of the medical profession – whom Wesley considered grasping and greedy – thought he was meddling in their terrain. However, his success prompted the establishment of the London Electrical Dispensary in 1793, paid for by public subscription. It treated more than 3,000 patients over the next decade. Portable devices treated thousands more.

The Prince of Electrical Joy

Another of the early electrotherapists, but in an entirely different mould, was the notorious 'Dr' James Graham, an entrepreneur who believed that 'electricity invigorates the whole body and remedies all physical defects'. James Graham was born in Edinburgh in 1745, the son of a saddler. He studied medicine at Edinburgh University, but never graduated, instead emigrating to the United States in his twenties, where, undeterred by his lack of formal qualifications, he set up as a doctor in Philadelphia.[3] There he was introduced to electricity by the striking lecture demonstrations of Ebenezer Kinnersley, a popularizer of Franklin's experiments, and he quickly became convinced that this new force was not only a universal panacea but also a means of making his own fortune. In 1774, he returned to England where he established a successful practice specializing in using electric shocks to treat a multitude of ills. One of his early patients was the eminent historian and political activist, Catherine Macaulay, who introduced him to her friends. Young, handsome and charming, he quickly became a society figure.

Emboldened by his success, in 1779 Graham opened the Temple Æsculapio Sacrum (the Temple of Health) in the elegant Royal Adelphi Terrace, facing the Thames. Two huge men clad in ostentatious liveries and hats trimmed with gold lace announced the grand opening by parading around London with handbills effusively extolling the delights of electricity and its effects on the human body, and proclaiming the enchantments that could be sampled at the Temple of

Health. The advertisers were nicknamed Gog and Magog, after the two legendary pagan giants who are reputed to have saved the city of London, and whose huge wooden effigies are housed in the London Guildhall today.

The Temple of Health both fascinated and outraged London society. It quickly attracted a rich and fashionable clientele who included the Prince of Wales and Georgiana, Duchess of Devonshire, who consulted Graham over her inability to conceive a son. For an entry fee of two shillings and sixpence, visitors could listen to lectures on health by Graham, enjoy fine music and marvel at the luxurious decorations, opulent furnishings and risqué paintings. Among the attractions were the scantily clad 'Goddesses of Youth and Health', one of whom was a beautiful sixteen-year-old called Emma Lyon. She was destined to marry Sir William Hamilton and later become the mistress of Lord Nelson.

However, the main attractions were the spectacular electrical delights – the 'magnetic thrones' and crackling electrical bath tubs. The pièce de résistance, exhibited only after a move to new premises in Pall Mall, was the electrifying and elaborately decorated Celestial Bed that was guaranteed to cure sterility and impotence. In Graham's words, 'The barren must certainly become more fruitful when they are powerfully agitated in the delights of love.' And some of that agitation was electrical. The Celestial Bed was twelve feet long and nine feet wide, supported by forty insulating pillars of brilliant glass, and had a mattress stuffed with hair from the tails of English stallions and sweet new wheat straw, mingled with balm, rose leaves and lavender flowers. Above it arched a vast dome festooned with flowers, live turtle doves, mechanical musicians and numerous statues, including one of Hymen, the god of marriage. But its novelty lay in the fact that sparkling electrical fire streamed from the torch Hymen held aloft and crackled across the headboard, illuminating the phrase 'Be Fruitful, Multiply and Replenish the Earth'. Reputedly, magnets surrounding the bed also filled 'the air with a magnetic fluid calculated to give the necessary degree of strength and exertion to the nerves.' It could be rented for the princely sum of £50 a night.

Like today's capital attractions, the Temple of Health also had a shop attached. It sold a variety of patent medicines: Imperial Electric Pills, for purifying the blood and precious bodily juices; Nervous Ætherial Balsam for decayed and worn-out constitutions; and phials of Electrical Ether to protect you against any kind of malignant or infectious disease, which was also reputed to have miraculous aphrodisiacal powers. Graham's treatises were also for sale.

The Temple of Health engendered a variety of reactions. The undiscriminating simply enjoyed the spectacle. Others were less impressed – the politician Horace Walpole dismissed it as the 'most impudent puppet-show of imposition I ever saw, and the mountebank himself the dullest of his profession. [. . .] A woman, invisible, warbled to clarinets on the stairs. The decorations are pretty and odd, and the apothecary, who comes up a trap-door (for no purpose, since he might as well come up the stairs), is a novelty. The electrical experiments are nothing at all singular, and a poor air-pump, that only bursts a bladder, pieces out the farce.'

After a brief period of notoriety, the Temple soon fell out of favour and closed in 1782. Graham had spent a considerable sum on the venture and was deep in debt. He escaped to Edinburgh, and in later life his behaviour became increasingly eccentric. He eschewed electricity and instead became convinced of the beneficial effects of warm mud baths, took to signing his letters 'Servant of the Lord, O.W.L.' (Oh, Wonderful Love) and once fasted for fifteen days wearing nothing but grass turf. He was arrested for indecency in 1794 and died soon after at the age of forty-nine.

The Tingle Factor

The exploits of James Graham and others like him engendered the idea that electrotherapy was mere quackery and it gradually fell into decline among the mainstream medical profession. Electricity continued to be used, however, in a variety of pseudo-therapeutic devices. It also enjoyed a vogue as an entertainment. Penny-slot machines flourished in Victorian amusement parks and on piers: grasp the handles of one of these machines and you were rewarded with a small electric current that produced a reputedly 'pleasurable' tingling sensation.

One of the best known of the Victorian medical equipment makers was Isaac Louis Pulvermacher & Co, whose patented electric chain bands were used for treating rheumatism, neuralgia and similar complaints. Advertisements claimed that philosophers, divines and eminent physicians, in all parts of the world, recommended them and that the effects were instant and agreeable. The cheapest cost five shillings, a not inconsiderable sum in those days. The Pulvermacher electric belt was made of copper or zinc and was dipped in vinegar before being applied. Some idea of what it looked like in use may be obtained from a description in *Madame Bovary* of the one used by the chemist Monsieur Homais. 'He was enthusiastic about hydro-electric Pulvermacher chains; he wore one himself, and when at night he took off his flannel vest, Madame Homais stood quite dazzled before the golden spiral beneath which he was hidden, and felt her ardour redouble for this man more bandaged than a Scythian and as splendid as one of the Magi.' Nor were the chains used solely by fictional characters. Charles Dickens, writing to the actress Mrs Bancroft, who had apparently recommended the device to him, said, 'As I shall be in town on Thursday, my troubling you to order the magic band would be quite unjustifiable. I will use your name in applying for it, and will report the result after a fair trial. Whether Mr. Pulvermacher succeeds or fails as to the Neuralgia, I shall always consider myself under an obligation to him for having indirectly procured me the pleasure of receiving a communication from you.'

A plethora of similar electrical machines proliferated during the Victorian age. They included devices for curing all manner of ailments, for stimulating muscles, and for invigorating the flagging male organ (it didn't work). A few are still in use today, such as the diathermy device that uses an electric current to heat the tissue or blood vessel to such a degree that blood coagulates. It is used routinely in surgical practice to cauterize wounds, as it is quicker and simpler than manually tying them off, or to destroy abnormal tissues, such as small polyps or cancer cells. In a bizarre twist, modified medical diathermy machines were used during World War II to interfere with the radio navigation system used by the Luftwaffe to pinpoint bombing targets over England, as they broadcast radio noise over a wide range of frequencies that jammed the German transmissions.

A Shock to the System

The sensational machines that caused a mild electric tingle subsequently morphed into those that gave a substantial electric shock. Electroconvulsive therapy (ECT) became the treatment of choice for patients with severe depression in the mid-twentieth century. It was introduced by the Italian physician Ugo Cerletti, who was looking for a new way to treat schizophrenia that had fewer side-effects and greater efficacy. At the time, it was believed that inducing a seizure in the patient could help the condition. This was usually achieved by administration of a drug (insulin, for example), but Cerletti was aware that an electric shock could induce epileptic convulsions in animals and wondered whether this might be an alternative approach for treating schizophrenia. Initially, he rejected the idea, as he was uncertain of the correct 'dose' of electricity to use, but he subsequently discovered that the slaughterhouse in Rome was using electric shocks to the head to stun pigs (prior to slitting their throats) and that if the animal was allowed to recover from the electric shock it seemed unharmed. He experimented to find the amount of current needed to stun a pig temporarily and then, in April 1938, he tested it on a human being.

The patient was a schizophrenic, who had been found wandering around the railway station suffering from delusions, hallucinations and confusion. He uttered incomprehensible gibberish. The first shock – which was given without anaesthesia – induced only an absence seizure, but the patient then spontaneously started to sing. Cerlatti suggested they repeat the experiment with a higher voltage, but before he could actually do so the patient suddenly sat bolt upright up in bed and cried out – in perfect Italian – 'Non una seconda! Mortifera!' ('Not another one! It will kill me!'). Undeterred, Cerlatti gave him a series of further shocks. After these, the patient seemed quiet and helpful, and was discharged from hospital professing he was cured. Subsequently, Cerlatti and his colleagues used ECT on hundreds of patients (and animals) and determined both the safest dose and which maladies it was most successful at treating. Its use quickly spread throughout the world.

In ECT, a brief electric shock is applied across the head. Its magnitude is sufficient to produce a severe seizure akin to that of a grand mal epileptic fit. Because the shock stimulates the part of the brain that controls the motor nerves, the muscles convulse and the limbs become rigid. Our limbs are controlled by two sets of opposing muscles, one of which contracts and the other of which relaxes when we make a movement. ECT stimulates both sets of muscles so that they contract simultaneously, making the limbs stiff and rigid. In the past, the convulsions could be so great that they even broke the patient's bones, but these days a muscle relaxant is administered to stop the muscle spasms and a general anaesthetic is also given. Usually patients are given several treatments, a few days apart.

Sylvia Plath describes her experience of electric shock therapy in a powerful way in several of her writings, such as *The Bell Jar* and 'The Hanging Man':

> By the roots of my hair some god got hold of me.
> I sizzled in his blue volts like a desert prophet
> The nights snapped out of sight like a lizard's eyelid:
> A world of bald white days in a shadeless socket.
> A vulturous boredom pinned me to this tree.

She was adamant she never wanted it again.

ECT was widely used in the 1950s and 1960s, but is much less common today, following the introduction of effective anti-depressant drugs. NICE, the UK's National Institute of Health and Clinical Excellence, advises that it should only be used to treat severe depression, severe mania or catatonia (muscle rigidity), and only when other treatments for these conditions have proved ineffective (at least a third of people with severe depression fail to respond to drugs). Ironically, given its history, ECT is no longer recommended as a treatment for schizophrenia.

Although its medicinal use is well established, ECT remains a controversial procedure. Whether it really works, and how well, is highly contentious. Some medical reports conclude that it is effective at relieving depression in the short term, whereas others have found no significant difference one month after therapy when compared to a placebo. In many cases the effects seem to be temporary and weak in the absence of concomitant drug treatment, and there is no evidence that it lowers the suicide rate of depressed patients. Yet some patients testify it has caused a dramatic improvement in their condition, ridding them of depression and enabling them to live a normal life once again – even to return to a high-profile career. This probably reflects the heterogeneity of the disease: clinical depression is unlikely to have a single cause.

Unfortunately, ECT is not without side-effects. All patients experience some short-term memory loss, presumably because the brain circuits associated with short-term memory storage are disrupted: indeed, this is one reason why short-term memories are believed to be stored as electrical signals. The accompanying amnesia has the singular advantage that most patients do not remember being given the shock. However some patients suffer permanent memory loss. Ernest Hemmingway, who had ECT in 1961, told his biographer: 'What is the sense of ruining my head and erasing my memory, which is my capital, and putting me out of business? It was a brilliant cure but we lost the patient.'

If the effectiveness of ECT is unclear, how it might work is even more uncertain. The massive shock affects the electrical activity of brain cells, causing them to fire at very high rates and generating an electrical storm similar to that seen in an epileptic seizure. One argument is that this leads to a massive release of chemical transmitters from nerve cells. As these chemicals control our moods, and the balance between them is thought to be disturbed in mental disorders, it is argued an increase in the concentration of certain transmitters may be responsible. But our brains are delicately balanced organs and what is needed is the right transmitter, at the right place, for the right amount of time – ideally without any increase in chemicals with opposing effects. How such fine adjustments can be achieved with something as crude as ECT is far from clear.

In the past ECT was exploited by some mental institutions to subdue troublesome patients. This abuse famously came into prominence in 1975 when Jack Nicholson starred in *One Flew Over the Cuckoo's Nest*, a film based on Kevin Kesey's novel of the same name, in which electric shock therapy was used by Big Nurse to instill fear in inmates and ensure they remained docile and acquiescent. It caused a sensation and led to a heated public debate on the use of ECT. Currently, one of the most controversial aspects of its use concerns whether it can be given without the patient's informed consent: this differs between countries, but it is legal in the UK and USA (although judicial consent may be necessary).

A Shocking End

A. S. Byatt's novel *Still Life* has a shocking ending – it concludes with the accidental death of the heroine, Stephanie Orton Porter, who is electrocuted by an ungrounded refrigerator in her own kitchen as she is trying to rescue a trapped sparrow. Byatt was once almost electrocuted in the same way herself, but was saved by her husband. As we all know, mains electricity can be dangerous. But how, exactly, does it kill you?

For a person to be electrocuted, sufficient current must flow to ground through their body to stop their heart, paralyse their respiratory muscles or severely damage their organs. The amount it takes to kill a person is small, only about 50 milliamps, which is why homes in many countries are protected by safety trip-switches that disconnect the electricity supply if they detect a dangerously high flow of current to ground. In the UK, such devices typically trip out in response to 30 milliamps of current flowing for about 30 milliseconds. The settings are even lower in the United States - around 5 milliamps for 30 milliseconds. This is because lower currents can also be dangerous; a current of 15 milliamps is enough to cause your muscles to contract so vigorously that you cannot let go of a live wire.

Electrocution was a relatively common occurrence when electricity was first introduced, and even experts were killed. As Hilaire Belloc succinctly put it,

> Some random touch – a hand's imprudent slip –
> The Terminals – flash – a sound like 'Zip'!
> A smell of burning fills the startled Air –
> The Electrician is no longer there!

It almost happened to me. Late one night, I was wiring up a high-voltage amplifier for use in my experiments. Being tired and careless, I accidentally touched the printed circuit board. I received a shock of almost 400 volts (DC) and ended up on the other side of the room, thoroughly shocked and badly frightened, with an arm that hurt all down its length. Yet it did not kill me. This is because it is the current that kills you rather than the voltage and, happily, in my case the current was very small, even though the voltage was very high. For the same reason, the shock produced by static electricity generators like the Van de Graaf machine used to produce spectacular 'lightning shows' in science museums, which may be millions of volts in magnitude, will not kill you – although it may make you jump and your hair stand on end – because the current is both momentary and very tiny.

But amps and volts are bound together in an eternal embrace by Ohm's Law, which states that volts equals amps times resistance. Volts are also dangerous because they drive the current through the body. Their ability to do so depends on the resistance they encounter: the higher the resistance, the more volts are needed to produce the same amount of current flow. Our skin has a certain resistance to an electric current and you probably won't feel anything if the voltage is less than 30 volts AC. If your skin is wet, however, its resistance falls and the threshold for electrocution is reduced. Thus both volts and amps are potentially lethal, depending on the magnitude and duration of the shock received and the resistance of your skin.

The War of the Currents

Electrocution is not always accidental. It is used as a means of capital punishment in several countries. The development of the electric chair is an extraordinary twisted tale of power, corruption and a desire to reduce suffering, and it went hand in hand with the choice of whether AC or DC electricity should be used to power the embryonic electric grid.

In electric circuits, current is defined as the flow of electrons through a conductor, such as a wire. If the current flows in one direction only it is known as direct current (DC), and if the direction of current flow alternates in a cyclical fashion it is known as alternating current (AC). Batteries supply direct current, but mains electricity is supplied as alternating current. The reason that mains electricity is supplied as AC current is because its magnitude can be easily increased or decreased using a transformer. This means that electricity can be carried at very high voltages (hundreds of thousands of volts) in overhead power lines but stepped down to normal household levels when it reaches your home. This is less feasible for DC current, which is one reason that AC electricity finally won out in the 'battle of the currents' and was adopted worldwide. In European countries AC current switches direction 50 times each second, whereas in the USA there are

60 cycles per second; the magnitude of the voltage supplied also differs, being 110/120 volts in the USA and 240 volts in Europe. The reason for these differences is largely historical.

Competing to 'electrify' New York City in the late 1880s were, on the one hand, Thomas Edison and, on the other hand, George Westinghouse and Nikola Tesla. Edison advocated the use of direct current and his Electric Light Company, in which his fortune was invested, was set up using DC power (at 110 volts). Unfortunately, power transmission at 110 volts is very inefficient and the voltage drops off so rapidly that each house would have to be within a mile or so of a power station – hence many power stations would have been needed to light New York. Increasing the thickness of the copper supply wires allowed the current to be increased without the wire melting but at a considerable economic cost. The other possibility would have been to make the transmission voltage more than 110 volts. However, this was not an option because there is no way to easily drop the high voltage to a lower one in a DC system. This fact also necessitated supplying different power lines for equipment that ran at different voltages. Imagine if different circuits were needed to power your washing machine, electric kettle and computer, and you will appreciate how inconvenient that would be.

Tesla contended that an AC system was a better option, and invented a means of generating and supplying AC power that was bought by Westinghouse's company. The advantage of this system was that electricity could be supplied at very high voltages in the distribution cables, so enabling it to be transmitted over long distances with less loss of power. It could then be stepped down to a lower (and safer) voltage at the house. Many fewer power stations were therefore needed, and only one power line to each home was required, as transformers could be stationed within the house to supply voltages at different levels.

The advantages of the Tesla/Westinghouse system quickly became obvious. Edison counter-attacked by arguing that AC current was highly dangerous and staged a series of gruesome public executions to prove it. In front of a large press audience, a stray cat or a puppy was placed on a sheet of tin and subjected to 1,000 volts from an AC generator – with predictable results. In one instance, the executioner almost electrocuted himself, being blown across the room by the shock, his body 'wrenched apart as though a great rough file had been pulled through it'. Seeking greater publicity, Edison electrocuted the elephant Topsy. It also did not escape his notice that a person would provide even better propaganda.

Old Sparky

On 6 August 1890, the New York prison authorities executed the condemned murderer William Kemmler using the electric chair. It was the first time it had been used and it was not a success. The first shock was too weak and failed to stop his breathing, so that the current had to be switched on for a second time. As the *New York Times* wrote, 'it was an awful spectacle [. . .] far worse than hanging [. . .] so terrible that words fail to convey the idea'.

The background to the story was that the state of New York had been endeavouring to find a means of capital punishment that was more humane than hanging. A member of the commission appointed to look into the matter, Dr Alfred Southwick, recalled having seen an intoxicated man die a quick and seemingly painless death after accidentally touching a live wire. The commission reported that electrocution was a possible solution, and on 1 January 1889 the state passed a law allowing the use of an 'electric chair' as a way of killing convicted criminals. There was just one small problem: no electric chair existed.

The state legislature did not specify whether AC or DC electricity should be used and left it to a committee to decide. Edison actively campaigned for AC current to be used, surmising the public would therefore not want it in their homes. He employed Harold Brown and Dr Fred Peterson to design an electric chair, and to carry out further public executions of dogs, calves and a horse using AC current, which generated considerable publicity. Given that Peterson was also on the government committee that selected the best method of electrocution, it is perhaps not surprising that AC current was finally chosen for the electric chair, thereby coincidentally stigmatizing it as too dangerous for domestic use.

Edison and Brown had to obtain the AC generator they needed by subterfuge, as the Westinghouse company refused to sell one to the prison for the purpose. Westinghouse, seeing his business interests endangered, protested that electrocution was a cruel and inhumane punishment and paid for Kemmler's appeal against the mode of execution. Edison was summoned as an expert witness for the state. The appeal was lost and Kemmler was electrocuted. However, the current was insufficient to cause instant death, and Kemmler was simply roasted. It was a far more agonizing way to die than hanging.

The electric chair works by stopping the heart or frying the brain. The prisoner is strapped into the chair and wires are attached to the skin by surface electrodes moistened with a conducting salt solution. A massive electric shock is applied that causes instant brain death, and subsequent bouts of current are used to ensure that the other organs are fatally damaged. Although it remains a legal form of execution in a number of states in the USA, it is rarely used today, lethal injection being the preferred means of capital punishment.

Edison may have been a great inventor and a brilliant businessman, but he was not without flaws, and his advocacy of the electric chair was far from glorious. It is ironic that he once boasted, 'I am proud of the fact that I never invented weapons to kill,' and that he supported non-violence towards animals. It also seems somewhat strange that Edison is fêted as the man who gave us universal electric light and power, given that it was the AC system that was finally adopted. A US hero, following Edison's funeral President Hoover requested that North Americans dim their lights for one minute as a tribute to his memory. By contrast, Tesla, who actually invented the national grid, is a largely forgotten genius.

Phasers on Stun

Society has often dreamt of a weapon that can temporarily incapacitate an individual, instantly stopping them in their tracks without causing pain or lasting harm. The 'stun' mode of the famous phasers used in the TV show *Star Trek* is just one of many fictional examples.

The latest real electric stun gun is the Taser. It works by stimulating your nerves so much that your muscles contract uncontrollably and you fall over, usually within two to three seconds. The Taser fires two small darts that are connected to a handheld gun by long fine cables. The darts pierce clothing and penetrate the skin, where they then serve as electrodes, conducting an electric current from the gun to your body. The stimulus causes your muscles to contract, incapacitating you for as long as the electric current continues to flow. It also hurts, because the electric current stimulates your pain nerve fibres. Indeed, it is not easy to say how it would be possible to make a device that stimulates your motor nerves, thus preventing movement, without also affecting your sensory nerves and causing pain. After allowing himself to be tasered in an effort to persuade the UK government to issue police officers with Tasers, Greater Manchester's Chief Constable said, 'I couldn't move, it hurt like hell. I wouldn't want to do that again.'

Tasers are now widely used by police forces to control violent people, or those suspected of being about to cause violence. Their use is not without controversy as a few people have died as a result of being tasered. Some people may simply be more susceptible to the electric shock, but others may have received a greater electric current because they happened to have a naturally lower skin resistance, or were wet when tasered.

Emotional Signals

As everyone knows, when you are very nervous you become hot and clammy and the palms of your hands get damp. Some people even break out in beads of sweat. This is because the brain responds to stress by increasing the activity of the sweat glands in your skin. The salty fluid they secrete decreases the resistance of your skin and this can easily be detected simply by seeing how readily a small electric current – so small that you don't feel it at all – passes through the skin. Skin resistance is highly sensitive to many emotions, including fear, anger and stress, and thus changes in skin resistance have been used to detect emotional changes in an individual. The psychoanalysts Carl Jung and Wilhelm Reich, for example, used it as a tool to help reveal their patients' emotional state.

Changes in skin resistance also form the basis of the polygraph lie detector, which measures the electrical resistance of the skin, the logic being that telling a lie will make you nervous. As might be expected, this system is not without flaws, as simply taking the test makes some people nervous while a hardened liar might not flinch.

Fascinatingly, the Church of Scientology uses similar technology in its 'E-meter', a pastoral counselling device that is stated to measure 'the electrical characteristics of the static field surrounding the body' and detect the subject's mental state. The E-meter became the subject of a major Food and Drug Administration investigation in 1963, following concerns that the Church of Scientology was using it to practise medicine without a licence, and that false claims were being made for its efficacy in treating various physical and mental illnesses. After prolonged litigation involving trial, appeal and retrial, the verdict was that the E-meter could only be used for religious counselling and that it should carry a warning stating it was not useful for the diagnosis, treatment or prevention of any disease. Not surprising, perhaps, since it is, after all, only a device for measuring skin resistance.

Mind Control

Electrical devices may have been used to kill, incapacitate or as instruments of torture and coercion, but they have also been used for good. Sometimes, as in the case of electroconvulsive therapy, the effects are controversial. But electrical devices are getting more sophisticated. Once we understand what the electrical activity of a cell or tissue looks like, it is often possible to provide an artificial stimulus that is an exact replica of the normal waveform to replace or correct a defective signal. Heart pacemakers have enabled many thousands of people to lead normal lives and implantable defibrillators have helped hundreds more.

The ability to control another person's brain remotely, to force them to behave in a specific way, is the stuff of nightmares, albeit perhaps the Pentagon's dream. Yet it is not impossible to control another creature's behaviour simply by stimulating the correct bit of the brain. José Manuel Rodriguez Delgado was sufficiently confident of this idea that in 1963 he stepped into a bullring in Cordoba in front of an aggressive fighting bull. As it charged towards him, Delgado stood his ground and calmly twiddled a button on a remote control device that sent a signal to a transceiver connected to an electrode implanted in the animal's brain. Electrical stimulation of the caudate nucleus stopped the bull in its tracks: it skidded to a halt within a few feet of the scientist.

Similarly, stimulating the brain of a fruit fly either with electric current or by photoactivation of light-sensitive ion channels can affect its course of action: as we saw in Chapter 11, it can cause a female fly to behave like a male one. Direct electrical stimulation of the human brain can have equally dramatic effects. Surgeons operating on the brain to remove a tumour, or tissue that is responsible for triggering an epileptic seizure, sometimes apply a small electric current to test that the tissue they are about to remove is not of vital importance for the patient. Such stimulation can provoke memories, sensations and even feelings of pleasure or fear. The right amount of current applied at the correct place can even be of immense therapeutic value. In some cases, the effects are so beneficial that electrodes have been permanently implanted in the patient's brain.

Parkinson's disease is a debilitating condition in which patients develop an involuntary tremor, muscle stiffness and difficulty in walking and talking. In some people the tremor is so bad that their arms windmill around wildly. Deep brain stimulation is now widely used to alleviate tremors that cannot be controlled by drugs. It involves electrically stimulating specific groups of nerve cells deep within the brain, usually in an area known as the subthalamic nucleus that is involved in controlling movement. The device used to do this is similar in concept to a cardiac pacemaker, and involves an electrode implanted in the brain connected by insulated wire to a small stimulating unit outside the body. A small hole is drilled in the skull and an electrode is inserted into the brain of an awake patient under local anaesthesia. The patient actively helps the surgeon decide whether the electrode is placed in the correct position by reporting what he feels when the stimulator is switched on and electrical pulses are applied to the brain. Once the electrode is in the correct position, the matchbox-sized stimulation unit is implanted under the skin, near the collarbone. Electrical impulses can then be sent from the stimulator to the electrode within the brain: usually continuous stimulation at a frequency of 150 pulses per second is used.

Deep brain stimulation suppresses the activity of the subthalamic nucleus of the brain. Quite how it does so is debated. One idea is that it stimulates the firing of inhibitory neurones that switch off overactive nerve cells: another is that it disrupts pathological brain rhythms. However it works, it is remarkably effective. Patients whose bodies are shaking uncontrollably appear normal the instant the device is switched on. Michael Holman, a journalist with the *Financial Times*, described it thus: 'It could not have been simpler or starker. At the touch of a button, the battery-operated stimulator implanted in my chest was turned off by the doctor in charge of my assessment. My tremor returned within seconds, steadily gathering force. In a couple of minutes I was shuddering and flopping hopelessly. Another touch of the button, and I was restored to my tremor-free state.'

Bionic Ears

Electricity has been used to power hearing aids for years, but these are no more than simple amplifiers that boost the sound. If the sensory cells in your ear are damaged you will be unable to hear, no matter how loud the stimulus. Normally, the hair cells in the cochlea of your inner ear sense sound signals and translate them into electrical impulses that are sent to the brain via the auditory nerve. In deaf people who have an auditory nerve that is at least partially intact it is possible to bypass the damaged hair cells and stimulate the auditory nerve directly. This is what cochlear implants do.

Currently, they come with both internal and external components, the former being implanted under the skin of the head and the other being worn behind the ear. The outer device, which is about the size of a small hearing aid, consists of a microphone, a speech processor and a transmitter. The microphone picks up sounds from the environment and converts them to electrical signals, the speech processor filters out background noise and the transmitter forwards the signals to a receiver mounted close to it, but inside the body. The receiver then sends the electrical signals to an array of tiny electrodes that lie alongside different regions of the auditory nerve. The array is introduced into one of the fluid-filled chambers of the cochlea during surgery, so that it lies close enough to the auditory nerve fibres to stimulate them externally.

The hair cells of the cochlea are arranged along its length according to the tones (frequencies) to which they are sensitive, with those that respond to high notes at one end and those responsive to bass notes at the other, much like the keys of a piano. The brain is able to discriminate pitch because different branches of the auditory nerve innervate hair cells that respond to different frequencies. Thus if a branch of the nerve is artificially stimulated the brain will detect it as a note of a particular pitch. The number of electrodes in a cochlear implant varies, with current devices having from sixteen to twenty-four. The more you have, the wider the range of frequencies you will be able to detect. This is why present devices cannot compare with the ear itself, which has more than 3,000 inner hair cells, enabling a far finer discrimination of pitch and the ability to hear musical composition.

Cochlear implants are currently only used in severely deaf people with damaged hair cells. They work best for adults who have lost their hearing and for very young children who are born deaf. There is a critical window for acquisition of language skills and it is important that children receive implants within that time period – two to six years is the typical age. The use of implants is still in its infancy and current devices do not provide people with entirely normal hearing: the British politician Jack Ashley once famously described it as sounding like a 'croaking Dalek with laryngitis'. It takes practice and training to understand the sounds that are heard, and it is particularly difficult for tonal languages like Mandarin in which pitch discrimination is essential. Nevertheless, many people who were once completely deaf can now hear, and even use the telephone. Understanding speech in noisy situations, like a busy restaurant or bar, however, remains a challenge.

Cochlear implants work only if a few auditory nerve fibres remain intact, which is not the case in all deaf individuals. To get round this problem, electrode arrays have been designed that are implanted into one or other of the hearing centres of the brain. While these work even less well than cochlear implants, they hold some promise for providing otherwise totally deaf individuals with a crude sense of hearing. Not all deaf people are interested in devices to help them hear, however. They resent the implicit implication that they are disabled, and prefer to rely on sign language, which enables them to communicate easily and fluently with one another.

Gripping Stuff

Every morning Christian Kandlbauer gets up, eats breakfast, climbs into his car and drives off to work. Nothing remarkable about that you may think, except that Christian lost both his arms at the age of seventeen in an accident. He now has two prosthetic arms: a conventional one and one that is controlled by his brain. The nerve that once controlled one of his lost limbs was surgically redirected to his chest and different branches of the nerve implanted in different muscle groups. Over time, new nerve terminals innervated the chest muscles so that now when Christian wishes to move his arm, his brain sends a signal down the nerve that excites the chest muscles. The tiny electrical impulses in the muscles are then picked up by an amplifier placed on the surface of his chest and translated into movements of his prosthetic arm. Prosthetic limbs controlled by thought alone are still in the development stage and Christian was one of the first people to be fitted with one.

Currently, most electrically powered artificial arms are controlled by electrical signals picked up from muscles of the residual limb, and the amputee must learn which muscles to contract to control the arm and then consciously do so. In general, such arms only allow a single movement – such as opening the fingers or rotating the wrist – at a time. They are also rather slow and not suitable for people who have lost the whole of their arm or leg. The more advanced type of limb, like the one that Christian possesses, enables far more complex movements that are controlled intuitively – as one patient said, 'I just think about moving my hand and elbow and they move.' But even these advanced artificial arms suffer from the drawback that there is no sensory feedback to indicate, for example, just how much force to apply to pick up an object – that needed to grasp a heavy jug might break a fragile egg. Bionic arms are also expensive and must be replaced every few years due to wear and tear. There is thus a pressing need to develop even better prostheses. As is so often the case with medical advances, war is the spur, and considerable investment in new prosthetic technologies has been stimulated by the large number of young US soldiers who lost one or more of their limbs while fighting in Iraq or Afghanistan.

A future dream is to enable the paralysed to walk by mimicking the pattern of electrical activity normally supplied to our limb muscles by our nerves. This is simple to state but extremely difficult to do, for walking is a highly complex task. It is not just that the artificial electrical signals must be supplied in the correct pattern and at the right rate to many different muscles, but that our movements are constantly adjusted by feedback from our limbs. Deep within our muscles lie sensors known as muscle spindles that detect the position of our limbs and the extent of muscle contraction. The information they supply is needed not only to enable us to walk properly, but also to cope with difficulties such as uneven ground or stairs. Thus some sort of feedback system may be essential if an artificial device is to send the correct electrical signals to the muscles.

Forward to the Future

The use of electrical devices in medicine is now routine. Deep brain stimulation has had transformative effects on the lives of people formerly incapacitated by the shakes, and its use in reducing severe depression is currently under investigation. Many people are able to lead normal lives due to cardiac pacemakers. Hearing aids have advanced into new territories. Prosthetic limbs are becoming ever more sophisticated. Devices to help the blind see and the paralysed walk are still in their infancy and there is a long way to go before commercial devices will be available, but there is no reason to think they will not exist eventually.

But it is unlikely to stop there. Functional magnetic resonance imaging (fMRI) can already be used to determine a person's answer to a yes–no question. In the future, with more sophisticated interpretation of brain scans, it may be possible to enable patients with 'locked-in' syndrome to communicate more fully. Whether it will be possible to read someone's mind, however, is a different matter. Current fMRI technology is massive, spatial and temporal resolution are limited, and how much can be interpreted from the signals produced remains controversial. Yet we should not forget that although the first ECG machine needed two rooms to house it, portable devices are now commonplace.

While pacemakers, deep brain stimulation and fMRI cause little comment, the idea of connecting your brain to a computer is far more startling. In one sense, many of us are already interfaced with our laptops and mobiles – although that connection is mediated via our eyes and fingertips rather than directly with our brain. But as I grow older, I would appreciate a more intimate connection. How wonderful to be able to access all my memories at will. To be prompted with the name of the person standing in front of me whom I taught twenty years ago and whose name now escapes me. To search the Web for information simply by thinking. Frightening though it at first appears to consider wiring your brain up to a computer, it is the nature of the connection that is the crux of the matter. Providing that it can be switched on or off at will, and that any information downloaded to a personal storage device (such as our brain) is both secure and under our own control, it seems likely that many of us will eventually succumb to its seductive lure. Thinking is, after all, faster than typing and reading.

But Mary Shelley's story has a long reach and first we will need to overcome our fear of the unknown, of monsters such as that created by Frankenstein. We will also need to find ways to legislate and regulate the use of such technology so that the poor are not disadvantaged. Furthermore, any such radical modification of our brains will need to be invisible (for humans prefer not to stand out in a crowd), and preferably it should be possible to remove it easily when desired. Today, we routinely enhance our senses with microscopes, telescopes and night-vision goggles, to name but a few examples, but we can take them off at the end of the day. Likewise, we have calculators and computers that immeasurably enhance our mental abilities and the

Internet serves as a vast, external collective memory, with far greater capacity and speed of recall than our brains and libraries. Indeed, many of us are rarely offline and our immediate response to an unknown question is usually to 'Google' it. It may be that some individuals will prefer to continue to access such electronic aids via their senses – their fingers, eyes and ears – rather than by a direct connection to the brain. But I for one would like a device that effortlessly stores and retrieves my personal memories, and it would obviously be invaluable for people suffering from memory loss caused by disorders such as Alzheimer's disease.

Artificial memory aids that plug directly into our brains are, of course, currently only science fiction. But science fiction often has a way of becoming science fact, and 100 years ago few would have imagined it would be possible to control a mechanical arm simply by thinking, or stop a charging bull with a signal to its brain. Perhaps in another hundred years such memory devices may exist. It is impossible to tell, but what I do know is that understanding how the body uses electricity, and how memories are laid down, stored and retrieved by the electrical circuits in our brain will be the key to their success.

Notes

Introduction: I Sing the Body Electric

1 For those who would like a more detailed explanation, it works like this. When the K_{ATP} channel is open, potassium (K^+) ions move through it, flowing out of the cell down their concentration gradient. Because K^+ is positively charged, its efflux makes the inside of the cell more negative. This negative membrane potential holds calcium channels closed, so preventing insulin secretion. When plasma glucose levels increase, more glucose is metabolized by the beta-cell. This generates a chemical called ATP, which binds to the K_{ATP} channel and causes it to shut. As a consequence, the membrane potential becomes less negative (as fewer K^+ ions leave the cell), which in turn triggers opening of the calcium channels. Calcium rushes into the cell and causes the insulin-containing secretory vesicles to fuse with the surface membrane and release their contents into the bloodstream.

1: The Age of Wonder

1 This quotation is often attributed to Galvani (see, for example, W. W. Atkinson, *Dynamic Thought* (Los Angeles: The Segnogram Publishing Company, 1906, p. 179) but this is not correct. It is not Galvani's style and he was not mocked in this way during his lifetime. The quotation was probably invented by the French astronomer Camille Flammarion as it appears in his book *L'inconnu et les problèmes psychiques* (Paris: Ernest Flammarion Editeur, 1862). I am indebted to Professor Marco Piccolino for this information.

2 Prometheus was sentenced by Zeus to have his liver torn out by an

eagle for eternity. His liver regenerated every night so his punishment was unceasing: it is fascinating that Zeus should have chosen the liver, as it is one of the organs most capable of regeneration.

3 Anne-Robert-Jacques Turgot's famous epigram on Franklin: 'He snatched lightning from the sky and the sceptre from the tyrant.'

2: Molecular Pores

1 Ribonucleic acid (RNA) and deoxyribonucleic acid (DNA). DNA is the molecular blueprint of our cells and RNA the messenger molecule that carries the information stored in DNA to the protein factories in the cell.

2 Rod MacKinnon won the Nobel prize in 2003, together with Peter Agre (whose story is told in Chapter 8).

3: Acting on Impulse

1 Hodgkin always attributed his success to chance and good fortune.

2 Huxley came from a distinguished family. He was the grandson of Thomas Huxley, Darwin's famous bulldog and a great promoter of the theory of evolution; and his half-brothers were the novelist Aldous Huxley and the biologist Julian Huxley. Hodgkin also came from an eminent academic family, many of whom were historians.

4: Mind the Gap

1 Despite its extreme virulence, botulinum toxin is easily destroyed by heating; however, the bacterial spores can survive a temperature of 100 °C for two hours.

2 It is ironic that 'gift' in German translates as 'poison', as what was indeed a gift to the West was poison for Hitler; the scientists he expelled helped the Allies win the war.

3 At the end of World War II, Feldberg was awarded a considerable sum of 'restitution money' from the German government. He generously used it to set up a fund for the furtherance of good relations between German and British scientists: each year it awards a handsome prize and financial support for one British scientist to visit Germany, and one German scientist to come to the UK.

6: *Les Poissons Trembleurs*

1 Socrates dryly replies that he resembles a torpedo only if the fish produces torpidity in itself as well as others, because the reason he baffles Meno is because he is confused himself.
2 This was the price in North Carolina, USA. No doubt they would have been more costly in the UK. A guinea is 21 shillings – one pound and 10 pence in current money.
3 Watts equals volts times amps.
4 Each ampulla consists of a small capsule that is connected to an opening on the surface of the skin by a jelly-filled electrically conductive canal. The receptor cells sit in the wall of the ampulla. They respond to a difference in potential between the canal lumen (which is continuous with the seawater) and the body's interior, which, in turn, stimulates electrical impulses in the nerve fibres innervating the ampulla. Cutting the nerves to the ampullae abolishes the shark's ability to sense weak electric fields, conclusively demonstrating that the ampullae serve as electroreceptor organs.

7: *The Heart of the Matter*

1 This was Herbert Gladstone (son of the more famous William), who was Home Secretary from 1905–10.
2 The mechanism is unclear.
3 He called his machine after Thanatos, the Greek god of death.

8: *Life and Death*

1 Most of the 150 to 200 litres filtered each day is absorbed in the upper part of the kidney tubule, by other kinds of water channel.

2 Raindrops do not (uselessly) trigger the trap because two hairs must be touched within 20 seconds of one another to produce a response.

9: *The Doors of Perception*

1 Aldous Huxley took the title of his famous book *The Doors of Perception*, about his experience of taking mescaline, from Blake's poem. The name of the 1960s pop group, *The Doors*, refers to Huxley's book, and through that back to Blake's poem. Blake in turn takes the idea from Plato, who famously remarked that we are like prisoners in a cave who see the outside world only as shadows on the wall, so that what we perceive is but an illusion of reality.

2 Too much vitamin A is very bad for you. The livers of some Antarctic mammals, such as polar bears and seals, contain toxic levels of vitamin A. The 1911–14 Australasian Antarctic expedition lost all their supplies and one of their party down a crevasse. The other two expedition members had to eat their huskies to try to stay alive – but Xavier Mertz died anyway, probably because he ate too much of the liver and developed fatal vitamin A poisoning.

3 People can see even with the lens removed, as only about 30 per cent of the focusing power of the eye comes from the lens. The cornea does the rest. Glasses can also help focus the light in those without a lens.

4 Later it was decided that colour blindness had nothing to do with it and that the driver had simply ignored the signal.

5 Because sound is produced by molecules colliding with one another to produce pressure waves, sound cannot occur in a vacuum. In space, no one can hear you scream and you cannot hear an explosion.

6 A similar effect is observed if you shout underwater – the sound travels less far than in air.

7 Of the band The Who.

8 Pickled artichokes don't work.

9 Contrary to what it says in the textbooks, the different types of taste buds are evenly distributed over the surface of the tongue.

10 The name derives from 'sauerstoff', the German name for 'acid', which is a solution containing a high concentration of hydrogen ions.

11 There are many more genes, but not all produce functional proteins.

12 Interestingly, in humans the same receptor detects the pungent oils in wasabi.

12: Shocking Treatment

1 For a video, see the Wikipedia entry on Topsy (elephant).

2 One of his machines can be seen at his house in London.

3 Practising as a physician without qualifications was not uncommon at that time.

Further Reading

Here are suggestions for further reading. I only include books and articles that are of general interest and readily accessible. For those who wish to know more, a more detailed bibliography may be found on my website.

Books

Ashcroft, Frances (2000), *Ion Channels and Disease*, San Diego, CA: Academic Press.

Bakken, Earl (1999), *One Man's Full Life*, Minneapolis, MI: Medtronic Inc.

Bryson, Bill (ed.) (2010), *Seeing Further*, London: Harper Press.

Darwin, Charles (1859), *On the Origin of Species*, London: John Murray.

Darwin, Charles (1875), *Insectivorous Plants*, London: John Murray.

Finger, Stanley and Marco Piccolino (2011), *The Shocking History of Electric Fishes*, Oxford: Oxford University Press.

Gregory, Richard (1997), *Eye and Brain: The Psychology of Seeing*, Oxford: Oxford University Press.

Hodgkin, Alan (1992), *Chance and Design: Reminiscences of Science in Peace and War*, Cambridge: Cambridge University Press.

Hofmann, Albert (1980), *LSD: My Problem Child*, New York: McGraw-Hill.

Holmes, Richard (2009), *The Age of Wonder: How the Romantic Generation Discovered the Beauty and Terror of Science*, London: Harper Press.

von Humboldt, Alexander ([1834] 1995), *Personal Narrative of a Journey to the Equinoctial Regions of the New Continent*, London: Penguin Books.

Huxley, Aldous (1954), *The Doors of Perception*, London: Chatto and Windus.

Ings, Simon (2007), *The Eye: A Natural History*, London: Bloomsbury Publishing.

Lane, Nick (2009), *Life Ascending: The Ten Great Inventions of Evolution*, London: Profile Books.

Lomas, Robert (1999), *The Man who Invented the Twentieth Century*, London: Headline Press.

Martin, Paul (2003), *Counting Sheep*, London: Flamingo.

Medawar, Jean and David Pyke (2001), *Hitler's Gift: Scientists who Fled Nazi Germany*, London: Piatkus.

The Oxford Companion to the Body (2001), Colin Blakemore and Sheila Jennett (eds.), Oxford: Oxford University Press.

The Oxford Companion to the Mind (2004), Richard Gregory (ed.), 2nd edn, Oxford: Oxford University Press.

Powers, Francis Gary and Curt Gentry (1971), *Operation Overflight*, London: Hodder & Stoughton.

de Quincey, Thomas ([1822], 1986), *Confessions of an English Opium Eater*, Oxford: Oxford University Press.

Quintilian (2002), *The Orator's Education*, trans. D. L. Russell, Oxford: Loeb Classical Library.

Raeburn, Paul (1995), *The Last Harvest*, New York: Simon and Schuster.

Rippon, Nicola (2009), *The Plot to Kill Lloyd George*, London: Wharncliffe Books.

Sacks, Oliver (1996), *The Island of the Colour-blind*. London: Picador.

Sacks, Oliver (1986) *The Man Who Mistook His Wife for a Hat*, London: Picador.

Shelley, Mary Wollstonecraft ([1818]), *Frankenstein: or, the Modern Prometheus*. Oxford: Oxford University Press.

Schmidt-Nielsen, Knut (1997), *Animal Physiology*, Cambridge: Cambridge University Press.

Streatfeild, Dominic (2001), *Cocaine: An Unauthorized Biography*, London: Virgin Publishing.

Syson, Lydia (2008), *Doctor of Love: James Graham and his Celestial Bed*. Richmond: Alma Books.

Wesley, John (1760), *Desideratum: Or, Electricity Made Plain and Useful. By a Lover of Mankind, and of Common Sense*, London: W. Flexney.

Articles

Feldberg, W. (1977), 'The early history of synaptic and neuromuscular transmission by acetylcholine: reminiscences of an eye-witness', in A. L. Hodgkin et al., *The Pursuit of Nature*, Cambridge: Cambridge University Press.

Harlow, J. M. (1848), 'Passage of an Iron Rod Through the Head', *Boston Medical and Surgical Journal*, vol. 39, pp. 389–93.

Hodgkin, A. L. (1977), 'Chance and design in electrophysiology: an informal account of certain experiments on nerve carried out between 1934 and 1952', in A. L. Hodgkin et al., *The Pursuit of Nature*.

Horgan, J. (2005), 'The forgotten era of brain chips', *Scientific American* (October 2005).

Kalmijn, A. J. (1971), 'The electric sense of sharks and rays', *Journal of Experimental Biology*, vol. 55, 371–83.

Kellaway, P. (1946), 'The part played by electric fish in the early history of bioelectricity and electrotherapy', *Bulletin of the History of Medicine*, vol. 20, pp. 112–37.

Krider, E. P. (2006), 'Benjamin Franklin and Lightning Rods', *Physics Today* (January 2006),

Lissmann, H. W. (1951), 'Continuous electrical signals from the tail of the fish Gymnarchus niloticus Cuv.', *Nature*, vol. 167, p. 201.

Loewi, O. (1960), 'Autobiographic sketch', *Perspectives in Biology and Medicine*, vol. 4, pp. 3–25.

Miesenböck, G. (2008), 'Neural light show: scientists use genetics to map and control brain functions', *Scientific American* (September 2008).

Quinton, P. (1999), 'Physiological basis of cystic fibrosis: a historical perspective', *Physiological Reviews*, vol. 79, S3–S22.

Acknowledgements

I could not have written this book without a great deal of help. I am grateful to many of my scientific colleagues for reading the chapters, helping ensure my facts are accurate and providing invaluable comments on content and style. I thank Richard Boyd, David Clapham, Nathan Denton, Carolina Lahmann, Chris Miller, Mike Sanguinetti and Walter Stühmer for bravely reading the whole book. For reading individual chapters or parts of chapters, I thank Jonathan Ashmore, Mike Bennett, Pietro Corsi, Keith Dorrington, Donald Edwards, Clive Ellory, Ian Forsythe, Uta Frith, Fiona Gribble, Andrew Halestrap, Judy Heiny, Edith Hummler, Peter Hunter, John Mollon, Keith Moore, Erwin Neher, Denis Noble, David Paterson, Marco Piccolino, Andy King, Geoffrey Raisman, Bernhard Rossier, Julian Schroeder, Paolo Tammaro, Tilly Tansey, Irene Tracy, Louise Upton and Gary Yellen. I thank Peter Brown for help with the Latin and Greek references, and for translating some of the original texts; Michaela Iberl for translating some German papers; Mathilde Lafond for help with the French translations; and Vivien Raisman for providing a modern translation of the Edwin Smith papyrus. Marco Piccolini and Bryan Ward-Perkins supplied historical information and advice, Andrew Forge provided the pictures of the hair cells, and Peter Atkins lent me his Galvani texts. Bruce Barker Benfield at the Bodleian Library very kindly showed me Mary Shelley's original manuscripts of *Frankenstein* and unearthed a letter from Percy Bysshe Shelley. I am particularly grateful to Peter and Karin Hunter for providing sanctuary in their beautiful home in New Zealand while I struggled with the first two chapters. Many friends and colleagues supplied me with interesting stories and I apologize to those whose stories or scientific research I was unable to include. Needless to say, any errors or infelicities that remain are my own.

Acknowledgements

There is an Italian proverb that states 'Se non è vero, è ben tro-
vato', which roughly translates as 'Even if it's not true, it is a good
story'. I have tried to ensure that the science is factually correct, but
it is more difficult to be as confident that the many historical stories
I tell are completely accurate or correctly attributed. In some cases,
names have been changed to protect an individual's identity.

I thank my friend and wonderful agent Felicity Bryan for encour-
aging me to write another book and never losing confidence that I
would eventually do so; my editors at Penguin, Helen Conford and
William Goodlad, for valuable comments, wise advice and listening
to my writer's agonies; Louisa Watson and Tertia Softley for their
careful copy-editing; and Patrick Loughran for his assistance with
the pictures. I am indebted to Ronan Mahon for the beautiful line
drawings. I am also grateful to my brother and sister for valuable
criticism and advice, and my fellow wordsmith, Chris Miller, for
helping to coin a few choice phrases. Most of all, I thank Tertia Softley and Iara Cury, who tracked down many obscure articles, winkled
many books out of the Bodleian Library and generally kept me
sane, as well as the members of my research team for their patience
and forbearance when I spent the weekends working on this book
instead of writing their papers, reading their theses or applying for
more grant money to fund our research.

Credits

Epigraph. Excerpt from Percy Bysshe Shelley letter to Ralph Wedgwood (15 Dec. 1810) held by University College. Reprinted with permission from the Master and Fellows of University College, Oxford.

p. 12. From *Novi profectus in historia electricitatis, post obitum auctoris* by Christian August Hausen, 1743. Reproduced by permission of the Royal Society.

p. 23. From *De viribus electricitatis in moto musculari commentarius* by Luigi Galvani, 1791. Reproduced by permission of the Royal Society.

p. 28. From *Essai théorique et éxperimental sur le galvanism* by Giovanni Aldini, 1804. Reproduced by permission of the Royal Society.

p. 33. Excerpt from *Discourses* by Jo Shapcott, published in *Discourses: Poems for the Royal Institution*, 2002. Reprinted by permission of the Royal Institution.

p. 57. Image generated and deposited into the public domain by the Electron Microscopy Facility at Trinity College.

p. 70. From *A Shoal of Fishes* by Hiroshige, 1832. Reproduced courtesy of the Caroline Black Print Collection, Art History Department, Connecticut College, New London, CT, USA.

p. 80. Brian Turner, excerpt from "Here, Bullet" from *Here, Bullet*. Copyright © 2005 by Brian Turner. Reprinted with the permission of The Permissions Company, Inc., on behalf of Alice James Books, www.alice jamesbooks.org. Reprinted by permission of Bloodaxe Books.

p. 118. From *Le tombeau de Ti*, II: *La chapelle*, Cairo: Institut français d'archéologie orientale 1953, pl. 85, by H. Wild. Reproduced courtesy of the Department of Ancient Egypt and Sudan, The British Museum.

p. 122, left. From *On the electricity excited by the mere contact of substances of different kinds* by Alessandro Volta. Philosophical Transactions of the Royal Society B vol 90, p. 403, 1800. Reproduced by permission of the Royal Society.

p. 122, right. From *Anatomical observations on the Torpedo* by John Hunter. Philosophical Transactions of the Royal Society B vol 63, p. 481, 1773. Reproduced by permission of the Royal Society.

p. 144. *Throwing a Dog's Heart-Beats on a Screen: The Scientific "Jimmy."* From the *Illustrated London News* (London, England), Saturday, May 22, 1909; pg. 735; Issue 3657. Courtesy of the Mary Evans Picture Library.

p. 162–63. Excerpt from "Goodness Gracious Me" by Herbert Kretzmer. Reprinted with permission of Berlin Associates Limited.

p. 211. Reproduced courtesy of Andrew Forge.

p. 232. *Cerebelo de paloma: celulas de Purkinje y granulares*. Reproduced by permission of Santiago Ramon y Cajal. Cajal Legacy. Instituto Cajal (CSIC), Madrid, Spain.

p. 285. Excerpt from "The Tender Place" from *Birthday Letters* by Ted Hughes, Copyright © 1998 by Ted Hughes. Reprinted by permission of Farrar, Straus and Giroux, LLC, and Faber and Faber Ltd.

p. 288. *Dispensing of medical electricity*. Oil painting by Edmund Bristow, 1824. Reproduced by permission of the Wellcome Library, London.

p. 294. Excerpt from "The Hanging Man" from *The Collected Poems of Sylvia Plath*, edited by Ted Hughes, Copyright © 1960, 1965, 1971, 1981 by the

Index

Note: page numbers in *italics* refer to figures,
and those with suffix 'n' refer to notes.

Index